Ego

The Fall of the Twin Towers and the Rise of an Enlightened Humanity

Peter Baumann
Michael W. Taft

16
EasyRead Large

RHYW

Copyright Page from the Original Book

NE PRESS
San Francisco, CA 94115

Published 2011

Jacket design by Brick Design, Inc.
Book design by Karen Polaski

Printed in Canada

Library of Congress Control Number: 2011910183

ISBN 978-1-60407-573-1

eBook ISBN 978-1-60407-656-1

10 9 8 7 6 5 4 3 2 1

This optimized ReadHowYouWant edition contains the complete, unabridged text of the original publisher's edition. Other aspects of the book may vary from the original edition.

Set in 16 pt. Verdana

ReadHowYouWant partners with publishers to provide books for ALL Kinds of Readers. For more information about Becoming A RHYW Registered Reader and to find more titles in your preferred format, visit:
www.readhowyouwant.com

TABLE OF CONTENTS

introduction i

PART ONE: The Prison of Feelings

chapter 1: evolution's unfinished product 3

chapter 2: what are emotions for? 18

chapter 3: the emotional nervous system 40

chapter 4: the pursuit of happiness 57

chapter 5: why do i care? 77

PART TWO: The Prison of Thoughts

chapter 6: becoming human 105

chapter 7: the conceptual revolution 128

chapter 8: prima donna 153

chapter 9: wake-up call 180

PART THREE: The Enlightenment Revolution

chapter 10: the organism is in charge 211

chapter 11: developing enlightenment 236

chapter 12: the rise of an enlightened humanity 265

peter baumann's acknowledgments 285

michael w. taft's acknowledgments 287

notes 288

references 300

about the authors and publisher 319

front cover flap 320

back cover flap 321

back cover material 322

Index 323

introduction

On the moonless night of May 2, 2011, specially cloaked Blackhawk helicopters slipped over the border between Afghanistan and Pakistan. Their target was Osama bin Laden, the most wanted terrorist in the world. After a fruitless decade spent hunting him in the wilds of Afghanistan, the U.S. military had found him living in a fortified mansion in the large city of Abbottabad, Pakistan.

The helicopters landed near the mansion, disgorging dozens of Navy SEALs into the darkness. The soldiers of SEAL Team Six ran upstairs, burst into bin Laden's room, and shot him twice: once in the chest and once above the left eye. Within twenty-four hours, they had buried his body at sea. It was the end of a terrorist mastermind, and the end of a megalomaniacal ego that believed itself to be the champion of Islam, the savior of Muslims worldwide, the spiritual heir to the prophet Mohammad, the defeater of the Soviet Army, and the scourge of America. While none of these grandiose visions was true, bin Laden was the man who inspired, organized, and financed the September 11 attacks on the United States—a turning point in world history.

The function of an individual human brain that creates and sells these stories about itself, that coordinates all the efforts to manifest that vision, is the ego. Until fifty thousand years ago, the ego didn't exist. But once it came into being as a function of the brain, the ego changed the world—both for good and for bad.

Fifty thousand years ago, the human species was rocked by a radical shift in cognitive ability. In the blink of an evolutionary eye, we went from being smart apes to being fully human, from living at war with the elements and each other to becoming civilized. Although outwardly our bodies did not change, inwardly our experience of the world transformed. This burst of cognitive power, called the conceptual revolution, paved the way for civilization to take hold. It allowed our attention to shift from a focus on the physical world to a focus on mental models so that we could plan more efficiently and imagine safer ways to live. But the most crucial new brain function that came on line during the conceptual revolution was the ego—the mental model of ourselves.

The function of the ego is to coordinate thoughts, emotions, body sensations, memories, and desires. It is an avatar for the whole self, a mental stand-in for our total organism that allows us to engage in complex behavior. Most humans call this chimera of brain software "me," firmly believing that this construction is actually who they are. Ego identification has been the defining human experience for millennia. But the age of the ego may be coming to a close.

We believe that the evolutionary process is moving the human species toward a watershed transition. A careful review of our evolutionary history shows that the development of the human brain has always gone hand in hand with expansion of technology and complexity of social networks. These three—brains, tools, social connections—form a feedback loop that continuously bootstraps humanity forward. The ability to conceptualize and create tools took us from ape to human. The ability to conceptualize ourselves and each other gave us civilization. The next cognitive step forward will be a massive expansion of conscious awareness: the ability we all have to witness our own usually subconscious brain function. Our Stone Age brain is still running the show, even in the twenty-first century, and it is riddled with cognitive biases, knee-jerk reactions, and an ego that traps us in a cramped, me-first, life-and-death struggle for survival in an environment in which that mind-set is out of date and out of tune. Increased conscious awareness will let us peek under the hood at our own deep psychology, understand our hidden motivations, and get a handle on these normally inaccessible drives and motivations. Most important, it will allow us to experience the ego directly as a function, rather than as "me": a profoundly liberating experience that will permanently change our relationship to our lives, ourselves, and others.

The evolution of the human race has not ended. Although the natural-selection pressures we face may no longer be biological, evolution continues, and our species changes for the better with each successive generation. IQs are steadily rising; human rights went from nonexistent to a dominant world issue in less than a hundred years; modern medicine formed itself in less than 150 years; democratic revolutions have been breaking out around the globe for more than two hundred years; businesses have steadily moved away from labor exploitation toward egalitarian teams; the Internet is turbocharging our access to new and different viewpoints, making us more aware and tolerant; and each generation relieves the culture from the pressure of yet another taboo, becoming more open about money, sex, relationships, illness, and mental and emotional difficulties. Instead of problems hiding behind closed doors, everything is more transparent and out in the open.

These changes signal that people are moving beyond the constricted, dominance-oriented, narrow self-interest of the ego and are starting to integrate themselves into the larger process of life as it evolves on the planet. Someday the sort of dangerous, delusional ego—so full of its own self-centered importance it can smash airliners into skyscrapers—may become a relic of our past.

PART ONE

The Prison of Feelings

chapter 1

evolution's unfinished product

On September 11, 2001, two jet airliners plunged into the Twin Towers of the World Trade Center, another hit the Pentagon, and a fourth crashed. In less than an hour, a group of just nineteen people had changed the course of world history. That such a small group of humans could have such a large effect on an entire planet of almost 7 billion is due in part to the fact that the terrorists had at their disposal all the energy resources, sophisticated tools, and complex interactive social structures of a modern technological society.

Imagine a genetically modern human living in the Paleolithic, about a hundred thousand years ago. In order to defend himself, he would have had a choice of two weapons: a rock or a stick. He would have crafted these by himself in a few minutes, chipping the rock into a hand axe or stripping the branch of twigs. A Boeing 767 aircraft contains 3.1 million parts, which come from more than eight hundred separate suppliers around the world. It is constructed in the largest building in the world at a plant in Everett, Washington, covering 4.3 million square feet, or about the size of about nineteen hundred family homes. It takes more than thirty thousand people working on three shifts to operate the factory, which has its own fire department, security force, medical clinic, electrical substations, and water treatment plant.

Even the most swiftly wielded stick has a limited impact. Our Paleolithic ancestor would only be able to hit one person at a time. His clumsy overhand hammering, which is our genetically programmed manner of striking, is notoriously inefficient. It might take many hits to do serious damage to an enemy.

A hunter-gatherer's only power would be the strength of his own arm. Al-Qaeda, on the other hand, had the imagination to turn a Boeing 767-200 into a highly explosive guided missile. Each airliner that hit the World Trade Center was loaded with about ten thousand gallons of jet-grade kerosene, enough to fill the gas tanks of five hundred minivans. The total energy of the kerosene was *1.3 trillion* BTUs, enough to power three hundred thousand houses for a month. The fire created by the explosive burning of this fuel was so intense that it weakened the Twin Towers' steel skeletons, causing them to collapse, killing several thousand people.

The hunter-gatherers of the Paleolithic lived in groups of up to fifty people. They were nomadic, roaming the landscape in search of food and erecting only temporary shelters. Each person required a minimum of a square mile to sustain themselves. Modern Manhattan has a population of about 1.5 million people living in just twenty-three square miles.

The Twin Towers had stood for thirty years and were about thirteen hundred feet tall, from which height people could see almost fifty miles in clear weather. On any given day, about fifty thousand people came to work in the buildings, with another two hundred thousand passing through for business or pleasure. The complex was so large that it had its own zip code (10048) and housed financial giants Morgan Stanley, Salomon Brothers, and Aon Corporation, as well as its own mall containing eighty stores including Sam Goody, The Limited, J. Crew, Banana Republic, Ann Taylor LOFT, plus restaurants, banks, and much more. It also contained a PATH train station that had been in service since 1909 and through which more than twenty-five thousand people passed every day to and from New Jersey. Atop the North Tower was a massive array of television, radio, and cell phone antennas, including a 360-foot-tall digital television mast.

The first plane, American Airlines Flight 11, hit the North Tower at 8:46a.m. By 8:50a.m. the firefighters of the Fire Department of New York (FDNY) had arrived and set up the first of many incident command posts to handle the fire. The original fire department in the city, back when it was New Amsterdam, was a bucket brigade organized in 1648; the Dutch brigades carried water in leather buckets made by the shoe cobblers of the colony. By 1731 the colony's new English rulers had sent two fire engines from London, which were pulled to the fire by volunteers, who then pumped the water by hand. Over the centuries this had grown into the FDNY, an emergency response organization of 11,600 firefighters and thirty-two hundred emergency technicians. The FDNY in 2001 used dozens of Seagrave Marauder II fire trucks with 500-horsepower engines, capable of pumping up to two thousand gallons of water per minute. In the first hours after the attack, 121 engine companies and sixty-two ladder companies rushed to the area.

The New York Police Department (NYPD), the largest in the world with around forty thousand officers and support staff, quickly sent emergency units and officers, as well as helicopters to report on conditions at the site. After the buildings collapsed, the NYPD took charge of evacuating thousands of civilians from Lower Manhattan. Boats came from all around the harbor, including commercial tugs, ferries, police and fire boats, local mariners, Coast Guard boats and cutters, navigating by radar through the thick smoke. They evacuated an estimated million people, who then began telling the world what they had experienced.

It is likely that humans of the Paleolithic era could speak only in a rudimentary way. Communication happened face to face, using gestures and sounds. There would be no written communication for another ninety thousand years, and news probably didn't travel more than a few miles at the most. On 9/11, humans around the globe gathered to watch the events in New York and Washington DC. By 11:54a.m. Space Imaging's Kronos—a commercial, high-resolution satellite traveling 426 miles above the earth at 17,500 miles per hour—had been redirected toward Manhattan and was taking pictures of the enormous cloud of smoke emanating from the cratered remains of the towers. With its one-meter resolution, details of the building debris and emergency vehicles were clearly visible. At the same time, a member of the Expedition Three crew aboard the International Space Station snapped photos of the New York region with a commercial digital camera from an altitude of 250 miles. People all over the world had almost instant access to these images. Since that day, communications about 9/11 have snowballed. An Internet search of "9/11" reveals over *132 million* results, and there are at least five hundred books on amazon.com in English on the subject. The September 11 Digital Archive contains more than forty thousand firsthand stories and fifteen thousand images.

International reaction to 9/11 ran the gamut from compassion and sympathy to victory celebrations. People in English-speaking nations offered their profound and sincere condolences and support. For example, thirty-five hundred miles away from New York, in London, the queen expressed "growing disbelief and total shock," and Ireland declared a national day of mourning. In Berlin two hundred thousand Germans marched through the streets to express solidarity with the US. The French newspaper *Le Monde* ran this headline: "We are all Americans." Leaders from all over the world expressed their sadness and solidarity. The majority of people in these countries thought of themselves as aligned with the US, or culturally similar. Identified, in some way, as "us." But that was not the whole picture. Although Yasser Arafat said he was "horrified" by the attacks, Palestinians on the street reacted differently. In Nablus people handed out candy to passersby in an atmosphere of celebration. And in Ein el-Hilweh and Rashidiyeh, two Palestinian refugee camps, there was ecstatic gunfire from AK-47s raised to the sky. Iraqi state television, then under the rule of Saddam Hussein, showed video of the Twin Towers collapsing, accompanied by the song "Down with America."

On September 20, 2001, the United States identified Osama bin Laden as the primary driver of the 9/11 attacks and presented the Taliban, rulers of Afghanistan, with an ultimatum to deliver al-Qaeda officials and cease support of terrorist operations in the country. The Taliban refused, and a coalition of allied Western forces began bombing al-Qaeda and Taliban targets throughout Afghanistan. To date, more than two thousand members of the Western military coalition have lost their lives fighting in Afghanistan. These deaths have included members of the US, Czech, Norwegian, South Korean, Turkish, and other militaries. A total of more than nineteen thousand Afghan troops and civilians, coalition troops, contractors, and journalists have lost their lives in Afghanistan. The cultural environment and foreign policy shift after 9/11 also laid the groundwork for the Iraq War, in which an approximate 890,000 Iraqi troops and civilians, US and coalition troops, contractors, and journalists have been killed.

Statistically, there was more violent death in the hunter-gatherer period; one out of three died at the hand of another. But mass murder did not exist; killing was not premeditated. Ancient hunter-gatherers did make warfare on each other, either individually or in small groups about the size of soccer teams. These affairs could be deadly but probably rarely lasted more than a single day. They fought to defend or acquire hunting grounds, foraging areas, water sources, and mates; they fought to fend off predators, including other humans. The stakes were obvious, physical, and immediate, and the choice was to kill or to die. The famous scene of conflict between two bands of proto-humans at the water hole in *2001: A Space Odyssey* depicts this sort of straightforward, survival-oriented struggle.

When necessities like food and water are contested, violence is the norm. But when our basic needs are met, we are less inclined toward fighting; thus violence has dramatically decreased over the millennia. A large segment of the human population is well fed and living a decent life, and fewer people, proportionate to the exploding population, are dying in fights over food. But the way we fight now has nothing to do with immediate survival needs. Instead our motivations come from elaborate mental concepts, and we plan in advance for future combat.

Osama bin Laden, a construction contractor by training and scion of the Islamic world's largest construction consortium, must have carefully considered how badly the impact and explosion of a projectile loaded with combustible fuel would damage the Twin Towers. Mohamed Atta, the leader of the Hamburg terror cell that perpetrated the acts, was an architect with a special interest in skyscrapers. Bin Laden would use the Hamburg cell to guide the missiles that would bring down these symbols of what they perceived as the American war on Islam. Atta and the eighteen other hijackers spent years in training and planning the attack through an international network of like-minded soldiers. They spent long stretches of time together discussing the conceptual elements of the doctrine they bound themselves to protect. Their months together in the militant training camps in Afghanistan, enduring the brutal physical regimen, all in a context of emotionally supercharged religious justification for murder, reshaped these men. It was not just words and images by which they influenced each other: their biology caused them to reflect each other's state of mind. Neurons in each individual brain attuned one body to the other, reflecting heart and respiration rate, level of muscle tension, and body reflexes in conjunction with their accompanying emotional expression. Buy-in to martyrdom was not just conceptual and emotional, but became a physical proposition.

Training someone to kill themselves for an idea requires subverting and overwhelming all of the survival mechanisms nature has baked into the substance of body and mind. The natural fear of death must be alleviated by a strong belief in a glorious afterlife. And the terrorist's conditioning, both mental and physical, created a profound identification: their very survival depended on destroying any threat to their way of life, any threat to the concepts—Islam, the Caliphate, their people, home—that had become indistinguishable from the sense of self. Al-Qaeda's ideology, strongly influenced by the writings of radical Sayyid Qutb, holds that the Muslim world is losing its spiritual foundation and regressing to the pre-Islamic state of superstition and evil known as *jahiliyya,* the "Days of Ignorance." Because modern, Western-style governments do not follow *sharia,* Islamic religious law, they are the enemy, pushing real Muslims away from their salvation. Qutb also passionately asserted that "world Jewry" was involved in conspiracies with evil forces (such as nationalism and modernity) to subvert and destroy Islam. According to Qutb, even most Muslims had drifted from the true religion and were actually apostates; they could be killed without sin. The 9/11 terrorists saw themselves as striking back at the forces that were threatening their most sacred principles, as well as the salvation and eternal life of their families, friends, and countries. Adding the sexual lure of seventy-two virgins to spiritual martyrdom for these young men was just the tip of an entire pyramid of imagery cemented in place with intense physical and emotional drives.

Although they had a very different effect, such conceptual and emotional triggers were present in al-Qaeda's victims as well. On September 10, 2001, fifty-eight-year-old Sonia Puopolo,[1] a resident of Dover, Massachusetts, changed her first-class ticket from Boston to Los Angeles five times. She insisted her family remain behind so that she could spend time alone with Dominic, her thirty-five-year-old son on the West Coast. A vivacious blond with a big smile, Sonia boarded American Airlines Flight 11, took her seat, 3J, behind and to the right of one of the hijackers. Just fifteen minutes into the flight, she would have seen a man stab a flight attendant. Blood sprayed everywhere and people were screaming. The man had what looked like an explosive device strapped to his stomach. A second attendant was also quickly killed. Daniel Lewin, a member of an elite Israeli special forces unit who happened to be on the flight, attempted to stop the hijacking, but was immediately cut down by Satam al-Suqami, a Saudi national and former law student. After that there was no resistance. One by one the passengers in first class were slaughtered as Sonia looked on in terror. Two flight attendants, Madeline Sweeney and Betty Ong, contact-

1 Name used with permisson of Sonia Puopolo's daughter

ed American Airlines and reported what was happening in the cabin.

One of the "muscle" hijackers, hefting a box cutter, killed the passenger in the aisle beside Sonia Puopolo. She was next. As she would have realized that her life was in immediate danger, the fear response—which human beings have inherited almost unchanged from the dinosaurs—must have exploded into action. Her adrenal glands secreted catecholamine hormones, kicking the sympathetic nervous system into high gear. Blood was shunted away from her internal organs and into the extremities. Her heart and breath rate, blood pressure, and glucose level skyrocketed, and the long muscle tissue of her legs plumped with oxygen-rich blood, all preparing her body for intense physical exertion. Her eyes would have dilated and her mouth would suddenly go dry.

Although we can hardly imagine a more personal threat, the fear she must have felt in that moment was anything but personal. It was more akin to a programmed mechanical sequence. Her brain, evolved on an African savanna to deal with the life-threatening presence of a lion or hyena, understood the mortal peril of the moment and responded in its predictable, prehistoric manner. Fear was not a decision or a concept; it was an orchestrated series of chemical responses and physical reactions that, in an ancient and irresistible way, had hijacked her body. But she was not in a Paleolithic forest facing a wild animal: she was strapped into the cushy seat of an airliner hurtling toward 1 WTC at 460 miles per hour, and there was nothing she could do; neither running nor fighting were options.

With her modern human brain, however, she could do something no other animal is capable of: escape into the imagination. She was a Catholic and believed in a benevolent, all-powerful god upon whom she could call for help. Shaking and pale, she may have closed her eyes and prayed for protection. This would have allowed her a moment of calm. She never had time to send messages to her family from the plane; it is likely she was among the first passengers killed.

The murder and mayhem of 9/11 was a tragedy on so many levels. That we have brains that can create things as complex and beautiful as airliners and skyscrapers is a near miracle of ingenuity and cooperation. And yet our minds can also dream up paranoid delusions and imagine taking revenge on persecutors, real or imagined. This capacity points to something very different in human behavior; something that has dramatically changed since our long-ago days as semi-intelligent social apes. Our powerful brains, with their capacity to create the wonders and beauty of the Taj Mahal, can also create the horrors and brutality of the holocaust.

***

Albert Einstein wrote of the human condition:

> *A human being is a part of a whole, called by us "universe," a part limited in time and space. He experiences himself, his thoughts and feelings as something separated from the rest—a kind of optical delusion of his consciousness. This delusion is a kind of prison for us, restricting us to our personal desires and to affection for a few persons nearest to us. Our task must be to free ourselves from this prison by widening our circle of compassion to embrace all living creatures and the whole of nature in its beauty.*

Trapped in a complex web of emotion and thought, we understand ourselves as a cluster of identifications with race, nationality, religion, political beliefs, age, gender, and profession. Alone, these concepts would be ephemeral, but they are underpinned by kneejerk emotional reactions, intense pleasure-and-pain conditioning in our bodies that resists any change or insight. Cobbled together, these concepts lock us in a sense of permanence and isolation, disconnected from the visceral common sense that would tell us at every moment just how wrong we are in our daily apprehension of reality.

Evolution acting through the mechanism of natural selection created the emotions that motivate and direct our behavior, as well as the imagination we use to model possible outcomes for that behavior. Over millions of years, these developed in our animal and proto-human ancestors into an extremely potent combination. As a system our thoughts and feelings have ratcheted our species up the escalator from harsh, brutal survival in the dirt to the comfort, cleanliness, and convenience of sipping tea on a transcontinental flight.

And yet, like any technology—our body/brain system represents a highly advanced biological technology—there is almost as much of a downside as there is an upside. The same imagination that allows us to build jet airliners can dream up a plan to crash them into skyscrapers full of people. Our religious feelings motivate us to feed and clothe the needy, or to kill nonbelievers. The same empathic emotions that allow us to care for our families and children can motivate us to annihilate anyone we think threatens our loved ones. This downside doesn't just drive international terrorists or even the murderer down the block. It drives the anxiety, depression, and alienation that plague us today.

Given that our bodies and brains, and therefore our thoughts and feelings, are the result of evolution, it is likely that evolution will also adapt in us a trait or capacity that provides a way out. Our genetic makeup is not written in stone, and is constantly changing. And although Einstein talks about taking individual responsibility to free ourselves from this prison—he was, after all, writing this quote in a letter to a rabbi—it may turn out not to be a personal matter at all. Perhaps it is more an issue of humanity, as a species, slowly marching toward an escape hatch.

The evolution of our species has not come to an end. Human beings are not a finished product, but instead a perpetually unfinished *process,* a moving target, and our current state, the human condition, is not the final word on the subject. Humanity is in motion as the wave of evolution continues to push us forward. The expansion of awareness that originally allowed us to become conscious of our thoughts and feelings is still under way. The rise in brainpower has not only created an explosion of skills—inventing tools, language, medicine, technology, civilization—it has at many times during the last two thousand years allowed some random outliers to glimpse something shocking: that who we *think* we are—our mental self-concept, or ego—is not actually what we are. Our self-concept is a symbol, an idea like any other. As evolution stumbles forward in its blind march of accidental brilliance, this radical insight that was once the province of a special few will slowly become the normal viewpoint: nothing special. The unfolding of the physical universe, the laws of nature, and evolution of life are generating the expanded perspective that will allow humanity to make the biggest prison break of all time—escaping the prison of ourselves.

chapter 2

what are emotions for?

If Richard Wajda had been on time to work, he would have died. He was a half hour late to pick up a friend's child, whom he then dropped off at the babysitter. He boarded the subway and headed for his job at the World Financial Center, a forty-story skyscraper across the street from the World Trade Center. The subway station was in the WTC, and while he was taking the sky bridge across to work, he heard a deafening explosion.

> *I looked up and saw flames coming from the Twin Towers. I immediately thought it was a bomb. Debris was falling everywhere as if it were the ticker tape parade for the Yankees. Suddenly I was hit in the head. I do not think it was a piece of the plane, I thought it might be office supplies because paper was floating down. But whatever hit me hurt me. I just turned around and ran across the street.*

Richard called his office, and a colleague told him that the news said to get off the street. "Well I was hysterical. As I was on the phone looking at the building, I saw a body fall from the building. Tears were running down my face." He called his mother next, who told him to run. When he got to Broadway, he turned back to look just as the second plane hit.

I did not see the second plane because I was on the opposite side of the building. All I saw was the burst of flames. Well, everyone was running as if Godzilla was chasing us.

I looked back then, and the smoke was following us. I kept on running—this time I was running along the riverside uptown to get far from the fire as possible. I was running with several strangers. A few of us then stopped in a nearby park to rest. It was far from the towers, but we were still able to see them, when then the first one fell.

The sound sent chills up my spine. I told the people to run, so we continued. We were so close to the United Nations, we were all in fear. Then the second tower fell. I called my friend, whose baby I had dropped off at the sitter, and told her what happened. She was able to see the destruction from Queens. I told her I was going to get the baby.

I then made it to the 59th street bridge to walk across to Queens. I ran across with thousands more. The fear of the bridge being hit was scary. I finally made it to the babysitter six hours later.

I made it home safe. I was hit with debris, I fell and was trampled. My legs are cut up, and

> *my back is in severe pain, and my feet are ripped up. But I am home and safe.*

In the months following the attacks, Richard made it his mission to travel America, visiting historic sites. But on July 4, 2002, in a *New York Times* interview, he said that he would be nowhere near any historic sites or big, public celebrations on Independence Day. "I am frightened by loud noises," he said. "Fireworks really scare me now."

Richard Wajda was only one of millions of people who experienced overwhelming fear that September day. Fear is the weapon of terrorists, and the 9/11 attacks accomplished that goal spectacularly. In the case of a life-threatening anthropogenic catastrophe of that magnitude, having a fear reaction makes good sense. It is how our brains are designed to respond. Shaped by the conditions of the ancient environment, they are the brains of the Stone Age, with an unsophisticated fear center equipped to react to the real dangers of the time: the woolly mammoth, the saber-toothed tiger, an attacking tribe, the deadly force of a jealous mate. Even though there were no skyscrapers or jet aircraft in the Paleolithic era, our brain can still recognize the danger of flames, explosions, and giant falling objects; it can see other humans dying; and it has a response that is perfectly engineered to keep us away from such dangers: *be afraid of that.* Although we are born with the fear response hardwired within us, the environmental triggers for the fear response can be expanded to adapt to our unique life conditions. Because the fear response can be learned, if a plane looks like it's going to hit a building now, chances are that we'll run before impact.

The trouble is that the brain applies such learned fear triggers in a general way. After you witness an event like this in person or on television, fear can become so pervasive that it turns into phobia: you might become afraid of flying in any aircraft ever again. Or of going up in a skyscraper. Or of going to New York City, or even any city at all. Or of people who look Muslim. Or, like Richard Wajda, of firecrackers.

***

The purpose of the nervous system is to guide the animal toward sources of energy and away from sources of damage. The nervous system motivates the animal to advance or retreat by generating *pleasure* or *pain.* Pleasure tells an organism that its behavior is correct, life affirming, repeatable. Pain signals an organism to stop its behavior and not to repeat it, because energy is being lost. And that is still the underlying motivation of all our behavior today. We need both pleasure and pain in order to function properly as organisms.

Although you might imagine that life without pain would be some kind of paradise, in reality it is a deadly business. Take the case of Roberto Salazar, a young boy who is one of seventeen people in the US with a genetic disorder known as "congenital insensitivity to pain with anhidrosis," or CIPA. Roberto is constantly laughing and smiling and is unusually happy. His mother couldn't believe what a good baby he was because he never cried about anything. As he grew a bit older, however, it became clear that something was very wrong. If he played outside in the heat, he couldn't sweat and would get heatstroke. He stopped eating altogether and had to have a tube inserted in his stomach. When he was two and a half, he broke his foot and walked around on it for days, oblivious. Like other children, Roberto went through teething, but unlike other children he gnawed on his tongue and lips to the point of mutilation. "If you could imagine when you bite your tongue how bad it hurts. At one point, you couldn't even distinguish that his tongue was his tongue," his mother said.

Roberto had to have most of his teeth surgically removed for his own protection. His hands stay wrapped most of the time for the same reason. Because he never feels fatigue, he is hyperactive, but because he never feels hungry and can't taste food, he despises eating. Roberto's entire family is on constant call, making sure he doesn't inadvertently injure himself. Without modern medicine and the twenty-four-hour care of his family, his life would have been very short. Most kids with this disorder die by the age of three. As of this writing, he's almost eight years old.

On the opposite end of the spectrum, there is the possibility of experiencing too much pleasure. Feeling good is an evolutionary signal that we have done something right for the survival of the organism. It's there to guide our behavior toward finding the right food, shelter, and mates to sustain and reproduce life. When the pleasure signal gets screwed up for some of us, this might mean overeating, substance abuse, or sex addiction.

A forty-eight-year-old woman had an electrode implanted in the right central thalamus area of her brain. It was connected to a button she could use to stimulate the electrode in order to control her chronic pain. She quickly discovered that pressing the button caused erotic sensations. The sensations were so intense that she developed a severe addiction to them. She used the button to self-stimulate so much that, just like any junkie, she began to neglect her hygiene and family commitments. She developed a wound on the end of her button-pressing finger, and she attempted to hack the device in order to get even stronger sensations. Struggling to regain her equilibrium, she asked the doctors to take the controller from her, but when they did she begged them to return it. After two years of this compulsion, she was a wreck, had become completely isolated, and was paralyzed by anxiety and lethargy. Nature put pleasure there to give us a little neurological cookie every time we do something it considers good or life affirming. When we subvert this guidance by short-circuiting the feedback loop, we respond just like a broken machine, a car that's lost its brakes, unable to stop before crashing.

The adaptive behavior guidance of pleasure and pain is layered in at every level of an organism, providing feedback about what is helpful and what is not. When conditions stay within a narrow window of homeostasis, you experience the body as feeling good (i.e., mild pleasure), the background hum of health and well-being. If sensors detect a deviation out of the prescribed range, you feel discomfort (i.e., mild pain) and are motivated toward a behavior that will bring things back into order. When you feel thirsty, yawn, or break out in a sweat, for example, it could be a response to this level of mild pleasure/pain feedback, the brain attempting to return your system to a comfortable homeostasis. All of this occurs without any conscious involvement on your part. It's a function of the default regulation package that comes with having a body, a perq of evolution that keeps your system running in top form.

When it comes to our own pleasure and pain, the sensations we experience in our own bodies, the body's impersonal nature is forgotten. Stub your toe in the dark on the way to the bathroom at three a.m. and a flash goes off in your head: it's *my* toe, *my* pain. Why did this have to happen to *me?* It's just nature telling you to watch out next time, nothing more.

The problems escalate when we carry that level of identification to *my* idea, *my* beliefs, *my* way of life. That's when the ego gets involved. The hijacker on Flight 11 who killed Sonia Puopolo was himself hijacked by his own ego's sense of identification with a deeply personal battle. And for her, the agony of her fatal wounds would have felt intensely personal: *my* body, *my* life. And yet even in that tragic situation, the nerve transmissions of pain were utterly mechanical, as were the nerve transmissions that release chemicals that produce a sense of peace at the moment of death. Nothing personal about it. Evolution designed the mechanism to withdraw and defend, and it kicks in whether that response makes any sense or not, whether it can be effective or not. Such responses have a mind of their own, one that does not feel the loss or the human tragedy. It's a coolly operating pleasure/pain neurochemical system out of which would evolve the ability to feel shock, horror, and sadness, as well as well-being and joy. It's the nervous system's—and the organism's—next step.

***

Understanding the purpose of Richard Wajda's fear is easy: it was his nervous system's reaction motivating him to get the hell away from a life-threatening situation. This same fear mechanism is present throughout the animal kingdom, in virtually every living thing on earth. Organisms must be able to recognize danger and respond adequately in order to survive. Evolution long ago weeded out ineffective strategies for self-protection and has held on to the behaviors that are effective at keeping us alive.

Consider how humans react to fear. If there is a loud noise, a person will immediately stop what he is doing and attempt to determine the source of the noise. He will then decide whether it is dangerous or not. If it turns out to be potentially dangerous, he will usually run away or hide. If neither of these works, and the dangerous thing—let's say it's a wolf—actually gets close enough to touch him, the person will scream and go into a series of programmed fight behaviors, including biting, scratching, kicking, squirming, and so forth. This pattern of fear behaviors is so similar worldwide, and occurs in response to so many different kinds of stimuli, that it suggests that these behaviors are largely genetically determined.

Even rats have the same response. In the presence of a cat, a rat will instantly stop what it was doing and orient to the cat. The rat will then either freeze or attempt to run. If neither of these behaviors is successful, and the cat corners the rat, the rat will vocalize (i.e., "scream") and attack the cat. The fact that this sequence of fear behaviors matches those of a human is no coincidence. According to NYU neuroscientist Joseph LeDoux, this fear sequence exists in virtually the same form in all vertebrates because it works to fend off attackers and promote survival, and so it has been conserved by evolution.

The corresponding sequence that occurs *inside* the body when danger threatens is also highly programmed. An animal reacts to a threat with a pattern of physiological changes that prepare it to execute the fear response as effectively as possible. The organism abruptly switches internal gears to become a turbocharged threat-response machine: increased heart rate and blood pressure, accelerated rate of breathing, leg muscles swelling with oxygenated blood. The fear machine is the same in birds, rats, rabbits, cats, dogs, monkeys, baboons, and people, as well as many other species.

All life needs energy to survive, and the core principle of the nervous system is to help the organism, through pleasure/pain responses, to advance toward energy opportunities and retreat from threats. Over time, as nervous systems developed from the jellyfish to the vertebrates, they gained a central processing hub: the brain. In addition to basic pleasure and pain, *emotions* came online as a second, more complex way to enhance the pleasure/pain response that increases the organism's chances of survival.

The hub of the fear system is the amygdala, a small almond-shaped structure in the brain. If you stimulate the amygdala in a human, as is sometimes done during epilepsy surgery, the person will report feelings of fear. If you stimulate the amygdala in a lizard, it will act as if a predator is present. Same area, same response. LeDoux says that the basic architecture of the fear system, with the amygdala at its center, was established at least as far back as the dinosaurs. "When it comes to detecting and responding to danger," he writes, "the brain just hasn't changed that much. In some ways we are emotional lizards."

Notice that this dinosaur part of our brain controls fear. It doesn't reveal our deepest feelings about life, art, politics, or religion, even though it often comes into play on such topics. Fear tells us what to avoid; anger tells us what to resist. Both of these emotions exist to keep us from dying, from getting hurt, from losing energy. They motivate and guide us to keep ourselves and our loved ones in good working condition. Natural selection carefully sculpted these feedback systems to help us survive, and they have done just that for millions of years.

The perspective that emotions are evolutionary feedback mechanisms, as opposed to being of psychological origin, is barely 150 years old and was first elaborated by none other than Charles Darwin. Already world renowned for his ideas about natural selection, Darwin's final book was a phenomenal bestseller called *The Expression of Emotions in Man and Animals.* He looked at how animals physically expressed emotions such as fear, surprise, anger, disgust, and affection and compared this to how humans expressed what are apparently the same emotions. His conclusion was that "The chief expressive actions, exhibited by man and by the lower animals, are now innate or inherited,—that is, have not been learnt by the individual." Our emotions are biological mechanisms that have evolved and been passed on genetically from our animal precursors.

For example, mammals express anger by baring their teeth. The expression is easily recognizable in, say, a growling dog or a hissing cat. And in something as closely related to a human as a chimpanzee, the expression is virtually identical to ours. People bare their teeth too, as if we are going to bite, but we no longer have the fangs to back it up. Humanity's ancestors had large canine teeth and were apt to bite each other. As social animals we need to signal our anger to others in some fashion, but the fact that we do it by baring fangs we no longer possess points directly to our animal origins. Darwin was correct to see this as an artifact of evolution. Once nature finds something that works, she sticks with it.

We spend the greater part of our lives trying to get rid of emotions we don't like and hold on to emotions we do like. Feeling sad? Watch a comedy. Angry? Go for a run. Something making you happy? Get as much of it as you can. Spend a day observing the reasons you do things and you'll find that mostly you are trying to change how you feel. Emotions rule us like puppeteers, and the more we reach up and try to pull our own strings, the more we find ourselves dancing to emotion's manipulations, in an endless cycle.

Our want/don't-want relationship with emotions, in which we are mostly trying to be in some other emotional state than the one we're in, precludes us from asking ourselves why we are feeling all these emotions in the first place. Not what we did to trigger a specific emotion, but why we have emotions to begin with. What is it about a human being that requires us to live like a fish on a hook, dragged around by sadness, fear, contentment, anger, jealousy, love? If we really are a part of the universe, the logical outcome of billions of years of physics and evolution, then emotions are part of that process as well. Why did we evolve emotions? What is it about them that impacts survival of the fittest? What are emotions *for?*

Emotions signal danger, push us to mate, or spur us to ward off an attacking enemy. Using piloerection—making hair, feathers, or whatever stand on end—one animal can signal another that it is dangerous, and maybe save itself from a damaging fight or getting eaten.

Think about the come-hither look your lover makes when he or she is feeling amorous. There is no misunderstanding what he or she is feeling, and what the expression of this feeling signals to you. In the same way, by expressing sexual attraction, animals can begin the mating process. And when one animal expresses fear, the rest of its flock knows to beware of danger. This is why we reflexively scream when something terrifies us. The scream itself doesn't help us in any way; instead the nervous system is using the body's warning system to signal the other people around us to watch out. The fact that the scream will happen even if there is nobody else around to hear it underscores the automatic, hardwired nature of the reaction. Natural selection put these reactions in us, and we exhibit them whether we want to or not.

***

In a letter to the editor in Nantucket's *Inquirer and Mirror* paper dated September 11, 2001, self-described liberal Steven Axelrod wrote the following:

> *I don't know what this makes anyone else feel, but I'll tell you what I'm feeling tonight: sheer red-eyed rage and fury. The amputation of the World Trade Center, the violation of my home town, the sheer senseless, blood and cant-soaked religion fuelled hatred of the act make me feel about the whole world of Islam what they have been feeling about us for decades. They want a religious war? I say give it to them. I say let them find out what happens when they awaken this sleeping giant. I say carpet bomb the whole Middle East—every one of those countries, with all the innocent people in them. This has to be a calamity for them, an act of God, a typhoon, a tidal wave, a rain of toads. They have to learn that they cannot let their lunatic fringe declare war on the most powerful country in the world*

> *because if they do we will reach over and crush them like the puny desert bugs they are.*

Alexrod quickly recanted, embarrassed by his expression of rage ("Perhaps I was insane when I wrote those things"), but the letter is a testimony to the heat of anger, the feeling in the body that arises when we feel wounded and afraid. The blood pressure jumps, the eyes narrow, the jaw sets, and the limbs fill with jet fuel. It took him days to calm down enough to consider what he was saying. Although we need the computer in our heads to assess a situation, it is the boiling blood in our bodies that tells us how we *feel* about it.

William James, the father of American psychology, said that these *physical manifestations* in the body are the salient part of emotions. Although we usually talk about emotions as mental events, for James they are physical. As he put it, "...bodily changes follow directly the perception of the exciting fact ... and feeling of the same changes as they occur IS the emotion." An emotion is made up of the bodily reactions to a stimulus. It's not that we feel afraid and then run; instead we find ourselves running—heart beating, lungs laboring, eyes wide—and realize that we are afraid.

Imagine being high up in the World Trade Center, quietly beginning your work day, when a plane slams into the building many stories below you. There is a huge bang, the entire structure jolts, and then you see the flames and smoke. Fire spreads to your office.

Windows are smashing as people jump from the building. Are you going to have a mental experience of "Hmm. I believe this situation is unsafe. Time to consider my escape options"? Or is a giant burst of adrenalin going to explode your body, causing your heart to pound out of your chest, your lungs to gasp for air, and your muscles to try to jump out of your skin? Fear will lift your body like a tornado and throw you toward any escape route it can find. All conscious thought will be left behind in the headlong drive to *live.* It is another layer of the nervous system's pleasure-and-pain guidance mechanism.

Such involuntary systems are essential for the human body to function at all. If you had to consciously monitor your blood pressure, glucose, and pH levels day and night, and adjust other systems on purpose to keep them in balance, this sort of physical housekeeping would take all your time and mental energy. It is, in fact, what a large part of your brain is doing 24/7, but well below the level of conscious awareness. Since evolution has built from the simple to the complex, from bacteria to human beings, this sort of bottom-up, automatic monitoring is the norm. Human brains operate on this principle, with the oldest, simplest part of the brain, the brain stem, taking care of all these housekeeping tasks for you. The frontal cortex, which is the most complex part of the brain, came the very last in evolution. This is the region that allows us to have a conscious awareness of an emotion.

Psychologists spent virtually the entire twentieth century attempting to prove that James was wrong, and they did find that his description was a little too extreme. There are some top-down (brain to body) aspects of emotion. The brain has to subconsciously evaluate the significance of a stimulus before there can be any reaction to it. For you to react to a bear, you have to first recognize that it is a bear, which is a function of the brain. You also have to remember that bears are dangerous and that you probably want to run away as quickly as possible, which is also a judgment ("appraisal" in psychology terminology) made by the brain in advance of any fear symptoms. It is this appraisal aspect of emotions, the idea that we determine what a stimulus means to us, that occupied the center stage of emotional research for decades. But much of appraisal happens below the level of awareness, outside the scope of the conscious mind. It is still because of our bodily reaction that we notice the appraisal has taken place and an emotional reaction has been triggered. As neurology is coming on strong in the twenty-first century, James's concept of the bottom-up (body to brain) nature of emotion is again gaining favor, at least in its more nuanced form.

Let's say you are having an anxiety attack and go to a psychiatrist who gives you a prescription for Valium. After you take the Valium, your anxiety has disappeared. You feel calm, relaxed, and peaceful.

Now let's say your partner is a triathlete who has pulled a muscle and she has a big competition coming up next week. She is having muscle spasms and goes to her sports medical professional, who gives her a muscle relaxant. After she takes the drug, her spasms subside. What's the connection? Both Valium and the muscle relaxant are benzodiazepines—the same drug in the same dose. The main effect of Valium is to relax the muscles, and this relieves anxiety because a large part of anxiety comes from feeling a state of tension in the body. The brain is getting feedback from muscle tone and using this data to tell you how you feel emotionally. And just as James said, if the brain notices that the muscles are tense, it interprets that as anxiety. So if you take a strong muscle relaxant and your body feels calm, warm, and soothed, it is very difficult to feel anxious. Cognitively you may realize that the external condition of your life is just as bad as it was before you took the Valium, but because of how the drug has changed the *condition of your body,* you feel that your situation just can't be that bad. It's the feeling in the body that shapes emotion.

Your evaluation of your emotional state changes when the bodily milieu changes. Give someone a dose of epinephrine (i.e., adrenalin) without their knowledge, and their emotional state—whatever it is—will become incredibly intensified, simply because of the drug's effect on their body. They believe that they are having a much more passionate emotional experience because their body is in a highly aroused state from the epinephrine.

Drugs aren't necessary to achieve effects like this. Studies have shown that if a depressed person produces a mechanical smile—just forces their facial muscles into a grin—after a half hour they report they don't feel so bad anymore. The body tells us how we feel.

***

Emotions were a somewhat neglected topic of biological science for most of the twentieth century until Paul Ekman took up the torch and began to research them in the 1970s. His initial goal was to support the theories of Margaret Mead, and nearly everybody else in the field, who postulated that emotions were socially constructed. We learn to express our emotions from our community, the anthropologists were arguing, and there was nothing evolutionary or genetic about it. Emotions were like languages: every culture had them but they were all different.

With this in mind, Ekman set off for the wilds of New Guinea, where he worked with the insulated Fore people, thinking he would prove that this foreign and isolated culture would have a unique system of emotional expression. They were far away from Europe, both in physical distance and cultural viewpoint, and it stood to reason that their emotional expressions would be very different.

One day Ekman was eating a can of American food, which the Fore people thought was repulsive. The expression of utter disgust on their faces was easy for him to understand. To his surprise, they also smiled when greeting friends and looked angry when somebody took something from them. Their expressions reached across the boundaries of culture and language and spoke directly to him. They were easy to understand.

Being a scientist, Ekman also conducted experiments with the Fore. He showed them photographs of Western people making various expressions and asked the Fore what emotion they thought was being expressed. They had no difficulty whatsoever identifying the feelings shown in the pictures.

Faced with this evidence, Ekman was forced to reverse his thesis about emotions. *Emotions and their expression were essentially the same in every culture around the world.* The only differences turned out to be in which emotions were frowned upon by a society or which feelings a culture attempted to conceal.

Our emotions and their expression derive from their animal precursors. They are part of the human system, a product of nature, not nurture. Think about the reaction of the Fore to Ekman's food. The fact that a tribesman thought that particular food was revolting was culturally determined—his people ate different food—but his reaction of disgust was something that even a cat or dog could probably recognize. He wrinkled his nose and set his mouth in a frown. Nothing could be a more obvious signal of *yuck.*

Disgust evolved to prevent animals from eating something toxic or diseased. The hunger drive is so strong that it requires an inhibition to counterbalance it and keep animals from making themselves sick with bad food. Think about the things that disgust us: urine, feces, other bodily waste products, rotting flesh and insects like maggots that are associated with it, and obvious symptoms of disease. The physical expression of disgust centers on "oral rejection," the sensation that you want to spit or vomit something out.

Like disgust, most emotional expressions continue to serve the same function in human beings as they did in their animal precursors. Facial expressions in particular have become central to human emotional communication. We need to know what other people are feeling, and to communicate what we are feeling to them.

Little Roberto Salazar cannot feel his own pain feedback; the steady hand of evolutionary guidance, developed over hundreds of millions of years, is missing for him. Due to a genetic glitch, the nervous system cannot give him its exquisitely honed signal to stop taking actions that damage the body. If we overindulge in pain killers, alcohol, or other substances that mask or skew the emotional responses that natural selection has braided into the human nervous system, it is like attaching a magnet to a compass—it can no longer provide the guidance it was built to give.

Our emotions are evolution's gift to us, a sort of neurochemical GPS navigation system for the human vehicle that provides continuous feedback, motivation, and direction for our actions. They are not the mirror of personality that we think they are, but rather a mechanism, a more sophisticated layer of pleasure and pain, meant to keep us on the right track. Our emotional expressions are direct descendants of those of our animal forebears, and it is an emotion's effect on our *body* that really matters. How they affect us personally is secondary to their role as a steering wheel for the organism. In fact, they are not personal. As Richard Wajda and so many others found out at the World Trade Center that September morning, emotional responses are not so much about feeling good or feeling bad, but about knowing when we are safe and when it's time to run for our lives.

chapter 3

the emotional nervous system

Howard Lutnick and his wife Allison were attending their son's first day of kindergarten when both their cell phones rang simultaneously, then suddenly cut off. Howard's driver told them that the World Trade Center had been hit by a plane. Lutnick was the CEO of Cantor Fitzgerald, a huge financial services firm with offices between the 101st and 105th floors of the North Tower (1WTC). There were over 650 employees, including his brother Gary, in those offices. Hopping in the car, Howard rushed to the site.

Seeing the smoldering gash high in the Tower, Howard jumped out and ran toward the people flooding out of the building, asking them, "What floor? What floor?" He was still trying to find someone who had escaped from anywhere higher than the ninety-first floor when he heard an apocalyptic roar.

"Like a combination of a jet engine in my ear, and metallic, like the Titanic hitting, an eerie sound," Lutnick said.

> *I ran, and there was this tornado following me, this giant plume of black smoke chasing me. I dived under an SUV. My glasses got ripped off. It was dead silent and pitch black. Every time I*

took a breath, I thought I'd choke, rubble coating my mouth and throat. I thought, I'm going to die. I can't believe I'm alive, and I'm going to die.

Lutnick made it to a pay phone, called his wife, and then found his way to the Greenwich Village home of Cantor lawyer Stephen Merkel, where they began making calls to determine who had escaped the offices. Slowly the realization dawned, that not one of the 658 Cantor employees in the office was alive.

Howard had lost his brother, many friends, and most of his company. The entire staff of Cantor Fitzgerald numbered only around nine hundred. Appearing on Connie Chung's show a few days after the attacks, Lutnick wept, saying,

> *...my brother, my brother was on the 103rd floor, he worked, he worked for me, he worked at Cantor, and he, he called my sister, just after the, just after the plane hit, and he told her that, he said that the smoke was pouring in. He was, he was stuck in a corner office, and there was no way out, and the smoke was coming in, and, he's not good, and things are not good, and he's not going to make it. And he just wanted to say that he loved her. And he wanted to say goodbye, and tell everyone that, that he loved them. Then the phone went, the phone went dead, so while I'm the head of the company, I'm trying to help my 700 employees who are missing their loved ones, I'm just, I'm just another one of them. I'm just another one of them.*

Later, fighting back tears, he said, "You got to live. But it's so sad. If you didn't have a greater purpose, you couldn't go on. There would be no point." Edie, Lutnick's older sister, also spoke of the loss of their brother: "If I only cry five times a day, that was a good day. Welcome to my world."

The crushing sadness of the loss of so many people in a tragedy like this is almost unimaginable. Yet we can all in some sense relate to the Lutnicks' pain because we have also felt the pain of loss, if on a smaller scale. Seen in the light of evolution, sadness is a response to failure, frustrated goals, or losses. Most significantly, it is associated with loss of attachment to a loved one. The basic function of an animal is to *gather energy* for life, and to *avoid losing energy.* Sadness is evolution's way of signaling that you have lost something that you invested a lot of energy in, something crucial to your survival in the wilderness. Extreme sorrow for a big loss, known as grief, is nature's way of warning us that if we lose our resources and partner we will likely die starving and alone in the wilderness: can't get much more depressing than that.

Positive bonding and social interaction release opiates—natural painkillers—in the brain. It feels *good* to like and be liked, to know your loved ones are safe. To the Stone Age brain, it means you are bonding with your tribe and you won't be kicked out and left alone on the savanna to die. This opiate effect, which has been studied in numerous laboratories, has led to the understanding that sadness in humans is actually a kind of opiate-withdrawal state, like a heroin addict having the drugs suddenly taken away. Some researchers are finding that grief, like other negative emotions, evolved from ancient pain mechanisms. If we lose a friend, the sorrow this evokes is meant to teach us to work harder to keep our friends next time. The Stone Age brain—our brain—experiences loss as a potentially catastrophic threat to survival. Any loss can be experienced by the ego as obliteration.

If sorrow is the response to loss, then happiness or joy functions as a positive reward for life-affirming energy outcomes. We feel happy when eating, feeding a child, and having sex: activities that are directly contributing to our genetic continuance. Happiness is an extension of pleasure, signaling that we have done something right. For people in the Paleolithic, having a big meal, finding a good camping spot, or getting over an illness was a big deal that nature needed to reward us for. We also got very happy about having loved ones and friends, because we required other people to survive; a person all alone in those days made a quick lunch for a predator.

Our neurological repertoire of emotional responses is capable of great simplicity and intensity, as well as great subtlety and finesse. We love the thick, rich soup of a nourishing human connection; the nervous system wants us to have it, and the ego closes the deal by creating an identification with other people—my friend, my lover, my group. If you try to have a close, intimate relationship with an alligator, you will probably be disappointed because reptiles are not capable of a wide range of emotional expression. Fear and aggression are about all the alligator is capable of. It cannot bond with you because there are no circuits for bonding in its brain.

Mammals, on the other hand, have a much more satisfying range of emotion. Think about how closely you can bond with your dog, the interplay of warmth and connection that is possible between you. You both have an "emotional brain," complete with intricate responses and circuits for bonding.

The human brain makes us by far the most emotionally responsive of all animals on earth, more even than any other mammal, because the fundamental wiring of the brain supports such responsiveness. The possibilities for bonding with another human are almost limitless and very deep.

The human brain is the result of a long process of trial and error. Its various modules were developed over the 3.5-billion-year history of life on earth and can be divided into three major parts. First, there is the *reptilian brain,* which is responsible for the mechanical, housekeeping functions of the body. It includes the brain stem and the cerebellum. The brain stem is the body's autopilot, making sure your heart stays beating, your lungs keep breathing, your guts digest food. It is the oldest part of the brain in terms of evolution. It is called the reptilian brain because you can find similar neurological equipment in your alligator friend.

The second part is the mammalian brain, or *limbic system.* This is much more advanced than the brain stem and constitutes the foundation of emotion.

The third part is the *neocortex,* which is the most recent, complex, and human part of the brain.

What we might call the emotional brain is mainly centered in the limbic system but has important components in the brain stem and neocortex as well. The core of the limbic system is the hypothalamus, which regulates the internal homeostasis of the body: things like hunger, thirst, sex drive, body temperature, sleep cycles. It does this by using hormones.

Let's say you have been exercising outdoors in the hot sun and you become thirsty. Receptor cells in the hypothalamus detect that the salt-to-water ratio in the blood is too high. So the hypothalamus releases a hormone that causes the pituitary gland to release ADH (antidiuretic hormone), which causes the kidneys to retain water, not releasing it as urine; this jacks the level of the water in the blood back up to the right level. It also makes you feel thirsty so you will go find something to drink. As you sit there, panting and hot, the hypothalamus also induces sweating to cool you off and a host of other changes to regulate your body back to a comfortable norm.

Because the hypothalamus is the gateway to controlling these bodily functions, the emotional structures in the brain seek to manipulate the hypothalamus in various ways. Our fear response wants to be able to keep the blood pressure high (which is good for running or fighting), and one way to do that is to have extra water in the blood. So if you have a close brush with a fatal car accident and are terrified, the emotional brain will stimulate the hypothalamus to release ADH to tell the kidneys to retain water. The emotional brain is promoting its agenda by manipulating the hypothalamus to control body functions.

The most important structure in the emotional brain is the amygdala, the core of the fight-or-flight reaction. It controls fear, anger, and other survival responses. It is the amygdala that recognizes the near-death danger of the car accident and commands the hypothalamus to release ADH. Because remembering dangerous stuff is so important to survival, the amygdala is also connected to a third structure, the hippocampus, which helps us form memories. The intensity of emotion we feel when encountering something determines how the memories of it are encoded in the brain. The amygdala is also involved in mating and other social interactions. It turns out that the larger your social network is, the larger your amygdala. Remembering people and your relationships with them is vital to your success as a human being. We are intensely social animals.

The most interesting part of the emotional system is the prefrontal cortex (PFC). The PFC is located right behind the eyes and is the newest structure in the brain, evolutionarily speaking. This supercomputer is responsible for abstract thought, symbols, prediction, mediating conflicting urges, deciding between right and wrong, and controlling emotional urges. Its planning and prediction capacity is there to keep us safe and help us gain energy. It imposes executive function on the other emotional structures, meaning that it is the source of top-down decision making, allowing for cognitive control of actions. It is crucial to our understanding of ourselves because this is the seat of conscious awareness, thoughts about emotions, and even thoughts about thinking.

The original purpose of the nervous system was to motivate the animal toward life-affirming behavior and away from life-harming behavior. The brain remains yoked to this original mission; it is a central processing station for information input and output at every level. Using emotions, the brain generates behaviors in an animal. It makes you feel angry so you will defend yourself. It makes you feel happy when you eat so you'll do it again. It makes you feel love so you will raise children. Emotions aim to regulate the homeostasis of the organism by inducing behaviors that gather energy and attempt to retain it. Emotions are an extension of the pleasure-and-pain binary code of the nervous system.

Feelings are feedback. They tell us what our ancient ancestors would find most life-affirming in a situation. Huge storm brewing on the horizon? Feel uneasy and seek shelter. Lion coming your way? Feel terror and run like hell. Attractive mate giving you attention? Feel attracted and bond. Child crying in hunger? Feel compassion and feed it. The list is endless.

This feedback can be a little out of date, like when a medical professional feels afraid alone in her home at night because no man is there to protect her, even though she has an alarm system and is capable of defending herself. Or when an executive manager feels lonely and scared because he has no close friends. Even though he is not going to die because there are no cohorts watching his back or helping him hunt, the deep, ancient substrata of his brain tell him through mute emotional feedback that he needs to find his tribe.

These emotional signals may no longer dovetail with our environment as perfectly as they once did, but they are the way our brain is built to function. The structures involved in emotion stretch from the prefrontal cortex, at the very front of the brain right behind the eyes, through the center of the brain and back to the cerebellum. The prefrontal cortex is the newest part of the brain, and the cerebellum one of the oldest. Thus our emotions connect our deepest, most ancient drives with our highest, most refined aspirations.

Yet fundamentally emotions don't function that differently from pleasure and pain. Although they are more nuanced and affect us in ways that feel much more personal—*I am angry; I am sad; I am in love*—they serve a function that is almost identical to pleasure and pain. They motivate us toward doing the right thing (adaptive behavior that increases our chances of survival) and away from doing the wrong thing (maladaptive behavior that could be detrimental). That is the bottom line.

Identifying with emotions in a way that makes them feel personal, unique, special, or precious misses the point of their actual purpose in our lives. They are not psychological: they are almost like the nervous system's gas pedal and brake. When we dig into the neuroscience and pick apart the neurons, getting at the wiring diagram, it is clear that the emotional system of a human being comprises areas that create drives and areas that suppress those drives. For example, the amygdala promotes seeking contact and sexuality, probably indiscriminately; it's whoopee time in the brain. The septal nuclei, on the other hand, inhibit this activity, effectively saying "Hang on there a minute," which allows for more selective attachment of partners. The two balance each other out.

Your brain structures evolved under very different environmental conditions; there were few loud noises, no people were around except your tribe, large moving things were dangerous animals, and your friends and family were a team that was necessary for everything from obtaining food to watching over you while you slept. When you, as a modern city dweller, are surrounded by loud noises like honking from lots of large, moving vehicles, everyone around you is a stranger, and you sit in your apartment alone, is it any wonder that you are filled with a nameless, constant, low-grade anxiety?

All these emotional flavors are part of our life-sustenance program, put in place by evolution long ago. While we may have some top-down control over a few aspects, it is clear that they want what they want when they want it because they want it. They are not there to make us happy. Not to reveal who we are as a person. If you were happy all the time, it would be a pathology, like the woman with an electrode in her head who stopped washing and caring for her family while punching the sexual stimulation button, or like Roberto Salazar, who had no sense of pain. You would have no motivation to do anything at all. Being too happy all the time would be the worst thing that could happen to you, for you, the people who love you, and the world in general. We require motivation to survive, and the constantly shifting symphony of emotions arising from the nervous system provides that motivation.

Remember the last time you fell completely in love? A brandnew, totally intense, gonna-die-without-them love? If you were head over heels, you probably felt like you were high on drugs all the time. Almost insane. When you were with your loved one, you felt butterflies in your stomach, you had difficulty concentrating, and had elevator feelings or the sense that you were flying. You wanted to be with your lover all the time, and if they left for even a brief period, you missed them and thought about them the whole time they were gone. You cared about them deeply, wanted them to be happy, were willing to do things to make them happy, and couldn't feel happy unless they did. Nothing in your life probably has felt more precious, more profound, or more important than your love. Nothing has felt so intimate and personal.

If love were only about reproducing the species, we wouldn't need it. The sex drive is plenty strong enough to take care of that. But human beings are in a bind with reproduction. A baby human takes a tremendous amount of effort and care to get on its feet. Human mothers are susceptible to miscarriages if they are too physically active in the last months of pregnancy. They need help to get enough food during that period. They are also very weak after giving birth, due to changes in the pelvis—brought about by our all-important upright, two-legged gait—that make giving birth difficult. The mechanics of walking upright made it necessary for women's pelvises to be much narrower than those of female chimpanzees and gorillas. This means that human babies are born many months earlier than would be ideal—almost natural preemies—in order for their large heads to be able to pass through narrow pelvis bones. Human infants are much more helpless than the infants of our ape cousins; human mothers are also more exhausted by the ordeal of giving birth.

So Paleolithic babies and mothers needed a man around to protect them and provide them with food during the vulnerable first months. While this sounds pretty sexist to our ears, remember that we are talking about the brutal conditions of prehistory. It was also in the man's genetic interest to stick around and make sure that the carrier of his genes, the baby, got a chance to grow up healthy and strong, which included having a mother to care for it. Human men and woman needed each other to raise a child.

But the man has a strong biological urge to impregnate other women, especially since he cannot father children for a while with the new mother. What could nature do to induce him into sticking around and taking care of a child? The woman, too, needs incentive. Other mates might see it as in their genetic interest to harm or neglect her children, so she needs the child's own father to stay if possible. Nature's answer is love.

Love is a powerful drug. The neurochemistry of love is so intense—flooding the brain with a cocktail of chemicals (dopamine, norepinephrine, and phenylethylamine) that causes an intense rush similar to amphetamines—that people actually become addicted to each other. When people have sex, bonding hormones are released: oxytocin in women and vasopressin in men. The more sex, the more bonding hormones, the greater the bond. Endorphins, the body's version of heroin, are also released during sex and other physical contact, which help in creating a dependency-like bond, cementing the relationship into the ego's sense of self. And researchers have found that people in love have reduced serotonin levels, just like people with obsessive-compulsive disorder, which causes thoughts about the object of their love to hijack their brain.

It's as if evolution threw everything in the chemist's closet at us to get us to stick together. This dirty trick of making us addicted to each other is so intense, it can become debilitating. Think of a teenager who is lovesick, moping around the house, unable to get out of bed, disheveled, and depressed. They have the classic symptoms of cold-turkey drug withdrawal because they cannot be near their loved one. And then there are sex addicts who go from fling to fling in order to get the drug high of love, destabilizing their lives in the process.

Love's notoriously brief duration is no accident either. Mothers and children are at their most vulnerable in the first two or three years after the child is born. That is the period when it is crucial they have the help of a man. Isn't it interesting that the passionate time of love, the Blue Lagoon period, lasts about the same length? Your partner hasn't changed, but now you can see their faults as faults rather than endearing quirks. The chemical cocktail has disappeared, right on schedule, shifting you into a new mode of behavior.

Even our supposedly sacred choice of the person we fall in love with is partially evolutionarily determined. We are sexually attracted to people who look the fittest and most virile or fertile. And women have a strong subconscious preference for certain types of sweat; yes, your man's B.O. is what you love about him most. It turns out that when parents have immune systems that are very different, their child's immune system is much hardier. Women given men's sweaty T-shirts to smell (in research situations) are most attracted to those from the men whose immune system is the most different from their own. They can most effectively taste it in his saliva. Women in experiments had no idea how they were arriving at their choices. It wasn't a conscious decision at all but the fine hand of evolution subconsciously manipulating them into having strong, healthy children.

***

The evolutionary purpose of love is all about raising children. Yet it feels somehow sacrilegious and dehumanizing to talk about emotions as if they are biological mechanisms. We are different from computers and robots, after all. And, to be sure, there is something important missing if we say that love is nothing but an expanded version of the nervous system's pleasure/pain responses. Stating that we know the biological purpose of emotions, and revealing their mechanistic nature, is not to say that that is *all* that they are. A machine will never know how it *feels* to be in love, from the inside out. A robot cannot know what it's like to cry at the loss of a marriage, to lose a job, or to feel rage at being attacked. We cannot even say that one person's experience of being in love is like another's. Our brains are extraordinarily complex, and each one is not only vastly complicated but utterly unique. What we feel inside is our own private experience.

But that does not stop us from coming to terms with the biological imperative, the physiological underpinning of that experience. Even if the way we feel is ours alone, the cause of that feeling can still be understood as the product of evolution, a simple stimulus-response formula. Denying or ignoring that fact will not protect humans from computers or prevent Skynet from triggering a nuclear apocalypse. It just robs us of an opportunity to understand human nature—with an emphasis on *nature*—more deeply.

Religion and philosophy have led a two-thousand-year smear campaign against emotions. Theologians said they led us into temptation, sin, and damnation. Philosophers felt the "passions" were something to be handled with the utmost care, lest they overwhelm, degrade, and destroy our life. Even science has mainly tried to avoid dealing with the emotions, feeling that they were too slippery and subjective to investigate with integrity.

Mass media have always fed our appetite to see the destructive aspect of the passions, but they also cater to its opposite, our cultural feeling that emotions express something special and authentic about us personally. The hunger to know how Lady Gaga "really feels" about gays in the military or what Jon Hamm "really feels" about having children reflects our cherished concept that emotions equal authenticity and connection. Virtually every movie or television show in history has celebrated the primacy of emotion over cold logic. Captain Kirk always chooses his illogical human feelings over Spock's machine-like reason. The machines always lose. When we read Richard Wajda's email, with all its grammatical errors, we feel as if it is authentic, and therefore trustworthy.

A good day for Edie Lutnick was crying only five times. Her brother Howard was emotionally breaking down in public. We understand that people in grief need to express it, get it out, and that eventually they will return to normal.

Understanding the evolutionary background of emotions is liberating. Armed with this knowledge, it is easier for us to allow them to get on with their functioning and to not take them so personally. As predictable biological responses, emotions are one of the least personal things about us.

Our internal, subjective experience of them may be exquisite, but it may also be nightmarish; if the nervous system that generates them is out of balance, emotions become every bit of the prison Einstein claimed they were. Our hearts get broken and our ego is crushed; our identity and life feel at risk. It's a very tough way to live, but it's the way we've been wired to survive.

chapter 4

the pursuit of happiness

A sanitation officer for the New York Police Department, Captain Charles "Chuckie" Diaz, was patrolling the base of the World Trade Center with his fellow officers on 9/11. As soon as the planes hit, the officers began helping guide people out of the area. His back was to the buildings when:

> *I then heard and felt something I will never forget as long as I live. It sounded much like a crashing freight train. It started with a rumble in the ground with turbulence all over and then the crashing grew louder and louder, each time throwing me to the floor again and again. The earth underneath me was buckling and my body slammed to the concrete pavement.*

The tower had come down right behind him. He was trapped under debris in total darkness. Gasping for air, choking in the dust cloud, he tried to escape.

> *All I could now see was a light waving back and forth and heard a terrified voice attached to it instructing me to follow this light. I tried to raise myself up, but the debris on my body was holding me down, but after a few attempts I was on my feet. I was now going towards the light and I could hear objects smashing to the pave-*

ment all around me and I moved as fast as I could.

Together with a group of dust-covered survivors, Diaz stumbled into the safe haven of a nearby Burger King. They were desperate to drink water, wash their faces, and get the dust out of their throats, but there was no water pressure. The Slurpee machine was still working, so they drank and washed with frozen slush. Spotting the flashing lights of an ambulance rolling past, Diaz hurried all the civilians outside and into the rescue vehicle. His right arm was hanging limply by his side, probably broken, but now was the time to attempt to locate his fellow officers who might be missing. He called for assistance on the NYPD radio and began the search.

After a few hollers, [his colleague] Sergeant Oloya appeared and we hugged each other happy to be alive.

They split up to look for others. Then the second tower collapsed and Diaz was again engulfed in a cloud of debris. Hurt and blinded, he stumbled around in a daze. A Federal Reserve officer grabbed Diaz's vest and pulled him across the street into a pizza shop.

Upon entering we were both given water to wash our eyes and clear our throats and it felt good to see again. Then someone handed me a telephone. I quickly dialed headquarters and gave Lieutenant Smith my location and requested help. I then said, "Do me a favor, if anything happens to me, tell my wife Barbara and my children Christopher and Valerie, I love them, and wish my daughter Happy Birthday for me." Then the phone went silent for a few emotional moments. The lump in my throat that I now felt and the tears in my eyes were not from the dust, and I was gasping for air.

Diaz arranged for two ambulances and, together with the officer, got all the people into the vehicles safely. He hitched his own ride in a police car down to the station at 125 Worth Street.

I was thankful as [an officer] helped me out of the vehicle and pointed me to the main entrance. Just as I entered the building, tears again rolled down my face. There in front of me, in plain view, was a sight that put a smile on my face ear to ear. As I walked faster and opened my arms to people, everyone else that was working with me at the World Trade Center did the same. We were all alive, we were all there. The emotion got the best of us, and we did not care.

One of the happiest experiences a person can have is to narrowly escape death. Winston Churchill wrote about the euphoria that came when somebody shot at him and missed. The movie *Groundhog Day* explores this idea by having Bill Murray's character get killed or commit suicide over and over, until eventually his bitter, cynical mood lifts and his life turns around.

In a 2009 *New York Times* article, author Tim Kreider wrote about nearly dying after getting stabbed in the neck. The knife narrowly missed his carotid artery but damaged the nerves that controlled his face. Despite his new lopsided smile, Kreider spent an entire year feeling tremendously good. His happiness was Teflon; nothing could bother him.

Emotions motivate and direct our behavior, and when we survive a situation in which we could have died, nature rewards us bountifully. Even if we logically had nothing to do with it (such as Churchill getting missed by a bullet), we nonetheless feel euphoria that can last for many months. Evolution dictates that we remember what we did to survive.

Some people make a lifestyle out of milking this response. By engaging in dangerous behavior such as extreme sports, skydiving, or base jumping, for a while they can tap into the euphoria of surviving.

Happiness in general is an emotional reward for energy-gaining or-saving behavior. When we get a raise, score the last fresh loaf of bread, or finally get to sleep after a long day, we feel happy. Passing our energy inheritance on to our descendants makes us happy too, as when we meet a new partner, have sex, or celebrate the birth of a new baby.

Having some small measure of material, social, and relationship success is probably necessary for basic happiness. As Maslow and others have pointed out, it is very difficult to feel good when starving, being homeless, or having no family or friends. The threat to life is just too immediate and intense. However, beyond this basic, reasonable level of comfort—enough food, money, and close relationships—there doesn't seem to be much difference in people's level of happiness.

Science long assumed there was a biological set point for human happiness: that our ability to be happy was determined by genetic factors. Research on identical twins raised in different families seemed to indicate something like a preset genetic level of happiness.

But this was before the discovery of neuroplasticity. The brain is not as unyielding and stiff as was assumed; rather, it is capable of large changes even in adulthood. David Lykken, behavioral geneticist and professor emeritus of psychology and psychiatry at the University of Minnesota, who did the most work with twin studies, feels that the genetically determined basis of happiness is at most 50 percent of the story. Another 10 percent comes from measurable factors in our lives such as relative wealth, status, health, relationship satisfaction, and so on. The final 40 percent, says Lykken, is the result of specific behaviors we do to feel better. The two most important things that affect this 40 percent are exercise—which boosts happiness greatly in the short term—and having a lot of human contact. Regular exercise and interaction with friends make us feel really good.

Many mental and physical health problems may be caused by a mismatch between our modern environment and our ancestral environment. For example, the Stone Age brain was programmed to live in tight little groups of people who knew each other their entire lives, which is probably the reason friends boost our happiness so reliably. We were not designed to live isolated lives in the midst of thousands of strangers, like a modern city dweller. Norwegian biologist Bjørn Grinde argues that we can only expect to experience happiness in ways that we were designed by evolution to live.

All animals since the beginning of time lived surrounded by nature. Our brains are hardwired at the deepest levels to respond to mountains, oceans, forests, fields, and other natural settings. The current environment of artificial light, loud sounds, and cement in all directions presents subtle threats to the unconscious mind. By making small changes to our lives, such as connecting with people and getting out in nature, we can live a bit closer to what our biological programming was designed for, and, according to Grinde, feel a bit calmer as a result.

While these little evolutionary hacks can raise our level of happiness a bit, in the long run they may not make that much difference. We soon grow used to whatever level of happiness or sadness we currently have, and look for a bigger boost elsewhere. But the search never ends. Trying to become permanently happy is a losing game.

The fact that we are always pushing ourselves to feel better is both an evolutionary necessity and an aspect of Einstein's prison of "thoughts and feelings."

We are driven to do things to try to feel good, yet we will never reach a state where we feel good for long. This is the evolutionary reality that self-help books do not want you to see, and frustration only intensifies as the ego's sense of entitlement comes into play in the face of loss of happiness. Even Tim Kreider, who was so elated after escaping a deadly neck wound, found his euphoria wearing off after a while. As the evolutionary feedback returned to normal, everyday anxieties and frustrations again possessed him.

***

In 2002 working class British teen Michael Carroll was on top of the world. He had just won a lottery worth £9.7 million (about $15 million) and exuberantly proclaimed himself King of the Chavs. A "chav" is a miscreant teenager given to drugs, crime, and violence: a punk. Carroll certainly fit the bill. As the nineteen-year-old garbage collector raised his champagne glass to toast his good fortune, he was wearing an electronic ankle monitor because of his many drunk and disorderly arrests. Yet posing for the cameras with a check the size of a door on his shoulder, he had every reason to be happy. He was young, rich, and ready for anything—except for what actually happened over the next eight years.

Carroll squandered his fortune on drugs, gambling, and thousands of prostitutes and lavished money on friends, family, and wild parties. Within a year of winning his fortune, he was seeing four prostitutes and smoking £2,000 worth of crack cocaine every day. In just eight years, he lost £1 million on dogs and horses, threw £1 million into his favorite soccer team, and lost £80,000 on a property deal in Dubai. Chunky gold jewelry and a massive Mercedes painted with "King of the Chavs" ate up millions more. He paid untold millions in property damage and lawyer fees, while his wife took their child and left him. A favorite pastime was driving around in his unmistakable Mercedes, sling-shotting ball bearings through the windshields of hundreds of cars, a diversion that added to his numerous arrests. Intruders killed his twelve dogs at a former mansion, which now stands empty and vandalized, the pool overflowing with garbage.

Today Carroll is broke, on welfare, and hoping to get back the job he had before it all started: as a garbage man. He told a newspaper that he had no regrets about his wasted fortune. "The party has ended and it's back to reality. I haven't got two pennies to rub together and that's the way I like it. I find it easier to live off £42 [welfare money] than a million."

Given the epic scale of his antisocial behavior, it might be easy to dismiss Carroll as a statistical outlier, the black swan of behavior problems, who couldn't be happy under any conditions. But despite his impressive lack of impulse control, Carroll's opinion of winning the lottery matches the norm. After an initial few months of intoxicating joy, most winners gradually sink back to whatever their original level of happiness was previous to their big win. Their wellbeing returns to its baseline like a homing pigeon. This is also true for other cases of happiness arising from particularly powerful positive experiences. It's not as if going out and attempting to survive another stabbing is an option. Yes, even happiness can be another iron bar in Einstein's prison.

Once an organism has been rewarded by its nervous system for good behavior, life needs to continue, and there's the rub. It still needs to find food and eat it, still needs to avoid predators and toxins. If the animal stayed euphoric forever, it would cease to benefit from the intelligence of emotional feedback. Wandering around in a state of euphoria, it might forget to eat and decide that predators were its best buddies. Although many spiritual and self-help programs purport to offer infinite, boundless ecstasy, it's a good thing that they rarely seem to deliver on this promise, as it would constitute a dangerous pathology. Emotional feedback is so important that we cannot live long without it. It would be very dangerous to be too happy all the time, like little Roberto Salazar, the boy who never feels pain.

***

Evolution has built into mammals an emotional regulator that resets the system after extreme highs and lows. Most of us believe that if we could just somehow manage to win the lottery, our problems would disappear, our anxieties and frustrations would evaporate, and true happiness would be ours. Yet this is not at all what happens.

Like the King of the Chavs, many lottery winners are ill equipped for the headaches of dealing with large amounts of money. But beyond this practical concern, something else occurs. The thrill of having a beautiful mansion and a sexy sports car wears off, and jetting to world capitals in first class becomes just another way to kill time. Movie stars, rock gods, and politicians are revealed to be regular human beings. Charlie Sheen, Lindsay Lohan, Winona Rider, Robert Downey Jr., Britney Spears, Ozzy Osbourne, Kurt Cobain, Rush Limbaugh: the list of people with supposedly super-fabulous lives who seek diversion in the most destructive ways goes on and on. Despite their best efforts to keep up the excitement, the lottery winner's glittery new life becomes rather mundane. While this may seem like a bummer, it is actually indicative of the system functioning properly, returning itself to a neutral baseline.

The technical term for this effect is "hedonic adaptation," also known as the hedonic treadmill, because the image of a person working hard but never getting anywhere fits the idea exactly. In terms of happiness, human beings are running just to stay in place. As psychologists Philip Brickman and Donald Campbell first described it:

> *As we pursue happiness, or any pleasant feeling for that matter, we are actually on a treadmill. As we acquire more of anything—money for instance or status—initially we feel good, but we soon get used to that level. What had previously seemed like a fortune now seems, well, not quite enough, and we need a little more to keep the same good feeling. An ever increasing amount of stuff is required to maintain the same level of satisfaction, or hedonic tone.*

Philip Brickman collaborated on another groundbreaking paper that examined the effects of winning the lottery or experiencing a crippling accident on people's happiness. Contrary to the common assumption of how these extreme positive and negative events would affect our lives, the researchers found that, after a short time, lottery winners weren't any happier than average, and people who had been tragically crippled in accidents weren't any less happy than average. Evolution and the nervous system want us to return to a neutral baseline in order to remain properly motivated and directed. If emotions stay too extreme for too long, a person will lose viability; they will be in danger of dying.

When we first read this concept, it can seem pretty unfair. Nobody likes the idea of having their happiness melt away. But the same is true for people who have suffered tremendously. In the rather literal, concrete viewpoint of evolved responses, accident victims who have lost limbs have done something "wrong": whatever behavior they engaged in that caused them to get into an accident was detrimental to their survival. So of course victims of such accidents feel terrible for a long time afterward. They are filled with guilt, shame, self-recrimination, anger, and depression—all part of nature's education plan.

Yet just as with positive emotions, these extreme negative emotions cannot last. No matter how intense or all-pervasive, they eventually begin to fade, and for the same reason as the positive emotions: the organism needs to return to baseline in order to regain viability. The neurological reset button gets pressed and, little by little, the life of the emotions returns to normal. Just as a tremendous stroke of luck comes to feel normal, so a horrible tragedy eventually becomes, well, kind of old news. Our emotional/energetic motivation system decides that it's time to move on.

Many of us live our lives as if there is an end state in which we will finally be happy. If we achieve a series of goals—get a good job, attach to the perfect mate, find worldly success of various sorts, achieve the needs of the ego, obtain the identity the ego consciously wishes for—our emotional state will settle into a kind of permanent happiness. Even if we think such a scenario is rationally possible in the external world, evolutionary psychology shows just how impossible such an internal emotional state must be.

No matter how perfect we make our lives, it is crucial to us as living organisms that our emotional state remains relatively at its baseline, neither too happy nor too sad. As a society we have doubled our income, added ten years to our life expectancy since the 1950s, and become smarter, yet our life experience is not any happier.

***

Staying relatively even-keeled may seem unfair, but it is what kept our ancestors alive and thriving, all the way back to the first bacteria. Another thing that seems unfair on the face of it is that there are more negative emotions than positive ones. The standard list of basic emotions is: happiness, anger, fear, sadness, and disgust. One positive emotion, four negative ones. Is this just more proof that nature wants you to be unhappy? Not at all.

First of all, in one way of looking at it, there are only two basic feelings: pleasant ones and unpleasant ones. Feeling good and feeling bad. Feeling good pushes us to do things that are good for us. Feeling bad induces us to avoid doing things that are bad for us. The ego identity seems to add a layer to this very simple system by giving us thoughts of *I want it; I don't want it; it should be,* or *it shouldn't be; I like this feeling of contentment; I don't like this feeling of anger,* but still, everything boils down to the same two things: the nervous system indicating pleasure and pain.

This advance/retreat, pleasure/pain model gives us a clearer view of what in general these feelings are trying to accomplish. Furthermore, in this way of looking at things, the list is perfectly balanced: it either feels good or feels bad. No problem.

When we categorize emotions beyond this simple binary system, we are making our best guesses at what the various subsets are. The taxonomy of emotion is still being hashed out. There are probably cases in which several English words describe the same emotion. Is envy really different from jealousy on a neurochemical level? There are other cases in which words don't exist in English to describe nuances that exist elsewhere. For example there's the German term, *Futterneid,* or "food envy," to describe the feeling you have when somebody else's meal looks and smells much better than your own. Or the untranslatable Portuguese word *saudade,* which refers to a particularly poignant, constant feeling of nostalgia, or deep longing for something that is gone, does not exist, or is unattainable. And when people of different ages and professions in a community feel a sense of face-to-face emotional connectedness in Japan, it is called *fureai.* There are literally thousands of these fine distinctions that may or may not be definable neurologically, but dividing them simply between pleasurable/painful or pleasant/unpleasant immediately simplifies the picture.

We really only need a few simple positive emotions and many more specific negative emotions. The reason for this imbalance is the "better safe than sorry" aspect of evolutionary fitness. That is, the cost for doing something wrong is very high—such as death—whereas the benefit for doing something right is simply continuing onward. This means that evolution has required organisms to develop much more specific and intense feedback around what *not to do* than what *to do.*

Let's take eating, for example. Nature rewards us for eating with the basic emotion of happiness. There are numerous other bodily feedback rewards, such as the pleasant sensation of fullness in the belly, blood glucose returning to the midrange or higher, and so forth. So we feel good emotionally in a basic, general way when we eat a meal, and that is enough, because the downside of not eating that meal is just that we will have to eat more later. It's not all that dangerous.

If, on the other hand, we eat something toxic, then it's extremely important that we spit it out immediately. It's not enough to just feel generally bad; we have to spit it out now. If we don't have a concrete behavior associated with this situation, the downside could be very high: sickness or death. Animals that just had a general negative feeling when they ate toxins are no longer with us. Those that developed a very specific emotional response of disgust lived. This is the better-safe-than-sorry reaction of evolution.

Another surprising factor that helps balance the emotional equation concerns anger. Is this a negative emotion because it feels bad, or a positive emotion because it feels good? The answer may not be as obvious as it seems. Even though most religions and philosophies have a special antipathy toward anger, saying it is the worst possible emotion, in terms of evolution and behavior the case is not so clear.

Anger can be considered a positive, optimistic state. Researchers have found that fear causes people to assess risk negatively, as we might expect. It makes them hesitant to take action. But anger has the opposite effect. Angry people are much less risk averse and predict *positive* outcomes for their actions. When irate, individuals are game to try something new, difficult, or dangerous; their choices closely resemble the decisions of happy people.

The rallying cry from the movie *Network,* "I'm mad as hell and I'm not going to take it anymore," taps into the motivating aspect of anger. By rousing an individual to action and lowering their risk aversion, anger can contribute to survival and well-being.

There is also a gender aspect to anger, or at least the aggressive behavior that can come from it. Women report feeling angry on average the same number of minutes per day as men do. However, the male brain's hormonal cocktail of testosterone, cortisol, and vasopressin favors taking aggressive action much more often than do women's hormones of estrogen and oxytocin. In fact, in men with a high testosterone level, anger and aggression feel particularly good. Evolution is telling them that it is time to demonstrate their dominance, which—if we were living in the Paleolithic—is a good way to get more food, more mating opportunities, higher status, larger hunting grounds, and a wealth of other positives. While this "bring it on" attitude might lead to very different outcomes in a modern society with police and lawyers, nobody has broken this news to the emotional structures in our brains. Our physiology has not yet caught up with our sociology.

Evolution is not done with human beings. Over the last fifty thousand years we have been undergoing a long process of domestication. The wild human of yesteryear, living in self-sufficient little packs in the field, is giving way to the tame human of today, living cheek by jowl in a hive of millions. This domestication is far from metaphoric: our physical bodies are changing in very noticeable ways, as is our behavior. Anger as a solution to problems is becoming less useful over time. Kicking the ass of enemies, which in the past saved your tribe, today lands you in prison. As civilization advances, the opportunities for win-win situations increase exponentially, thus cooperation becomes a far better strategy than confrontation.

This is one of the many cases in which our evolutionary heritage is slightly out of date and no longer exactly matches our life circumstances. From this perspective, it is likely the case that anger in humans is gradually lessening, or at least the violent behavior that comes from it is. Although it may not always seem like it, we live in the least violent age human beings have ever experienced. Our wild, untamed reactions are slowly being bred out of us by civilization.

***

In Oscar Wilde's classic novel *The Picture of Dorian Gray,* the title character mistreats his lover so cruelly that she commits suicide. Immediately afterward he spends a pleasant night out at the opera. His friends are horrified at his coldness, but he says, "A man who is master of himself can end a sorrow as easily as he can invent a pleasure. I don't want to be at the mercy of my emotions. I want to use them, to enjoy them, and to dominate them." Wilde intends his readers to be repelled by Gray's insensitivity. Yet in a less extreme tone, Gray's comment might typify our cultural attitude toward emotions.

We do not want to suffer through negative emotions, and we want to induce positive emotions at will. As a society, we distrust and reject our own physical, emotional responses. Christianity, for example, has had a long history of repressing emotion and demonizing the physical body. The best example of this is the teaching of the Seven Deadly Sins. According to Catholic tradition, some sins are relatively minor ("venal") and can be washed away by confession. But the "cardinal" sins are much more dangerous. If a person commits a cardinal sin and dies before confessing it, their soul will go to Hell. These extra-powerful seven sins are lust, gluttony, greed, sloth, wrath, envy, and pride. Each one is supposedly so evil, so powerful for shepherding souls toward destruction, that Satan has given seven special demons the job of tempting humans in these sins.

From the perspective of evolution, however, the Seven Deadly Sins are nothing more than a partial list of basic drives and emotions. The drives for sex, food, energy (money), and sleep are there, as well as the emotions of anger, pride, and envy. These fundamental emotional and physical responses of the human body cannot be suppressed for long. For example, the desire to reproduce is fundamental to life. It is hardwired into the system as the most deep and most ancient force, existing even in bacteria.

In his 2010 visit to the United Kingdom, Pope Benedict expressed his "shock and sadness" at the sex abuse scandal that was battering the Roman Catholic Church, saying that it was "difficult to understand the perversion that had been revealed" in the priesthood. Yet it is obvious that denying the sex drive leads to the worst sort of misbehavior. In the case of the Church, it also compels the need to hide, deny, and lie about what is, after all, our evolutionary inheritance.

Nature equipped us with these drives and emotions to help us survive. Those who lacked these qualities perished and those who possessed them lived, reproduced, and bequeathed their successful genes to us, their descendants. Such deeply physical processes must find a physical expression, and if they are not allowed a healthy and open one, they will settle for a sick and twisted one.

Teachings like the Seven Deadly Sins relegate our natural, emotional life to the demonic, something morally wrong and dangerous that we have to take special spiritual steps to avoid. Yet how is lust—the drive toward reproduction—any different in terms of evolutionary responses from sleeping at night or eating food? It isn't.

Animals need to reproduce, and nature has ruthlessly eliminated any animals that did not possess a strong drive to engage in sex. Whether positive or negative, pleasant or unpleasant, emotions are evolutionary signals that arise from the nervous system according to their own ancient logic. We may be able to change them to our liking in the short run, and we certainly don't act on every impulse, but our emotional life has its own natural balance to which it will always return. This may sound tragic or deterministic, but it actually points the way toward freedom.

chapter 5

why do i care?

Rosemarie O'Keefe was driving across the Brooklyn Bridge on the way to her job as the commissioner of community assistance, when she received a call from Mayor Giuliani's scheduler. She was told a plane had hit the World Trade Center and that she should go down there immediately. She and her team stayed for three months, until it was time to switch gears for a new mission. She formed the 9/11 Family Assistance Center in a small auditorium in the NYU hospital. As they were finishing the setup, she stepped outside for a breath of fresh air.

> *I went out the front door and there was a thousand people lined up around the block. I said, "How do they know we're here?" There was a thousand people. They wanted to come in ... I told them that the moment we had lists (of who had been killed in the WTC) that we would share it. And that's what these families wanted. They wanted lists. We ended up in that hospital lobby auditorium for a few hours and we outgrew it.*

Her center, which helped the grieving families of the 9/11 victims, outgrew every facility it occupied until finally it moved to Pier 94, a space of some 130,000 square feet. There she set up sixty different units to help people find out everything they could about those they'd lost. The center had a DNA identification room; immigration services and help for foreign nationals; Safe Horizons, an assistance group for victims of violence; and psychological counseling. But Rosemarie knew it was not enough. She arranged for the families to go to Ground Zero to get some sense of connection with their deceased loved ones, as well as closure. They took the families down the East River, one by one, on a ferryboat.

> *Staff took three trips a day. Most of the families were hysterical. They gave them such comfort and such strength. They took them down on the water. They circled the cove. And when they got to the site where the buildings were, we gave each family member flowers, and they had prayer cards. And they would go and they would place these items in the most respectful and dignified manner at the site.*
>
> *Most family members came with extended family. Somebody. But this one lady had nobody. She was traveling alone. And she was going to go on the ferry down to the site. And so I put the pet therapy dog and his owner on the ferry. And that dog instinctively knew that she was alone. And he went over to her and he put his head in her lap for the entire trip. And then when we got off the boat, he walked with her to the site where she could put the flowers down.*

Rosemarie realized that even visiting Ground Zero was not enough for these survivors who had lost so much. They needed some kind of token or remembrance. She and her team designed a round, wooden memorial with a brass nameplate. They presented these to the families one at a time in a private room, with an official escort and an American flag. One gentleman held up the memorial, saying, "This is my wife. This is what I have of my wife. And this is what I will treasure the rest of my life."

Why did Rosemarie O'Keefe and her team work so hard to help the families of 9/11 victims feel better? What was it that drove so many firemen to ignore their own safety and run upstairs *into* the burning Twin Towers? Human beings can be remarkably kind to each other: everything from giving someone the time of day to sacrificing our lives for our fellows. Most people spend a majority of their time cooperating with other people to make the world turn round. The news often ignores this, instead focusing on all the ways we can be awful to each other. We take our altruistic, cooperative behavior for granted, which is why it is not news. Yet in terms of evolution, it is truly amazing.

We are a social species, like wolves, orcas, bees, ants. Humans evolved in small, tightly knit clans in which everyone helped each other. A group of twenty-five or thirty people cooperating to hunt meat, gather fruits and vegetables, and care for children is the perfect life our genes imagine for us. Privacy was unknown, and under these intensely social conditions, humans and other, earlier social species evolved special social regulators—drives, motivations, communication channels, and feedback loops between individuals—that helped the group as a whole survive. Nature equipped our precursors with a set of emotions tailored to regulate the group toward success. We are internally and biologically motivated to cooperate. Built to get along.

In prehistoric times, the success of the individual depended entirely upon the success of the group. In nature there was no such thing as a human being living alone. If we grew fed up with our Paleolithic clan and wandered off on our own, we would quickly find ourselves in big trouble. Who would feed us if we were hurt and couldn't hunt? How could we make all the tools, clothing, and weapons we needed while also having the time to hunt and gather? How could we defend ourselves against predators like lions and wolves without a large group of friends? Who would stand watch for animals and enemies while we slept? In the pretechnological world, being alone meant we would end up as food for somebody else almost overnight.

Social behavior didn't start with human beings. Prey species like fish, birds, and antelope come together in flocks, schools, and herds of hundreds of individuals, moving together as a coordinated group. By quickly changing directions like a perfectly synchronized *corps de ballet,* the group confuses predators, who need to choose and focus on a single animal weak and slow enough to catch. The herd makes that task difficult. Aeronautical studies of flocks of geese show that their iconic V formations make it easier for each bird to fly. Just as one Tour de France cyclist drafting another uses 30 percent less energy, the geese benefit from one strong leader plowing through the wind. Wolves are mediocre hunters on their own, subsisting on rabbits and carrion, but when they hunt together as a pack they often bring down moose, caribou, and other large animals.

The energy expended caring for each other in a group is more than balanced by the energy saved in driving off predators and the energy gained in more abundant food opportunities. Cooperation works, and so natural selection favors it.

Cooperation makes obvious evolutionary sense, yet actual altruism—where one animal helps another at the expense of its own energy or even its life—was difficult for science to explain. Behaving altruistically costs an animal energy, thereby reducing its fitness. Meanwhile the animal being helped gets its fitness increased. Over many generations, it seems like those with the gene for altruism should thrive less often, because they compromise their own survival, whereas those that take advantage of the other's altruism would flourish. The net result, the theory seemed to say, was that altruism would always be stamped out by evolution. Nice individuals would be weeded out by natural selection, replaced by mooches who sucked up all the help they could get. Not only did that not happen, but evolution has also selected for another ego trait: to be invested in being perceived as a good person. Unless you're crazy, you want to show others that you are a nice person.

The secret is in our genes. All individual lives eventually pass away, but our genes are immortal. Copied from parent to offspring, they travel down the generations, and it is genes that drive, and ultimately benefit from, evolution. If an animal is unsuccessful, its genes are not passed along. Successful animals pass along their genes to posterity. We are the result of an unbroken chain of evolutionary successes that stretches back to the first life nearly 3.5 billion years ago.

Seen in this way, at least half the answer to altruism becomes clear: by helping those carrying the same genes, we are helping our genes thrive. Because we share a large number of genes with our parents and siblings, we all benefit from family altruism.

This idea, called "kin selection," finds its most powerful expression in insects such as bees. Why should a worker bee give her life for her hive and queen? As a result of insect biology, bees in a hive share a very high number of the same genes: 75 percent. By helping her sisters, the worker bee ensures that a large number of her own genes get passed along. Because her genes continue on, her behavior, which is encoded in the genes, can also continue on.

But we don't just demonstrate compassion toward family members; game theory explains why. Most games we play are zero-sum games, meaning that there must be a winner and a loser. But there are some games that are not zero-sum, in which *both* players can win. In soccer all eleven members of a side are working together for a *common* victory. If their side wins, they all win. Win-win-win-win-win-win...

There are many examples of non–zero sum situations in biology. Birds will cry out when they spot a predator. Yes, this reveals the bird's location to the predator, but it also warns the rest of the flock, allowing them to escape. Vampire bat young will die if they don't get fresh blood often, so the adults help each other out by regurgitating their meals, thus spreading the good luck around and feeding those young bats who would otherwise starve. As civilization has advanced over the last ten thousand years, the opportunities for non–zero sum interactions between human individuals has grown exponentially. Thus our biology actually pushes us not only to help ourselves but to help others with whom we share a common good. Compassion is part of our biological toolkit—it's the same impulse that caused rescue workers to ignore the risk of death in the burning Twin Towers in order to save lives. They were attempting to bolster the numbers of the tribe.

As social animals, we don't like to lose tribal members—especially not our most valuable hunters—so when something goes wrong, compassion and regret play a dual role in keeping us all together.

In December 2009, golf star Tiger Woods crashed his black Cadillac Escalade into a fire hydrant and then a tree. A few days earlier, *The National Enquirer* had published a story from a woman claiming to be Woods's mistress. On top of everything, Woods's wife smashed the car windows with a golf club. The circumstances raised some difficult questions; so did Woods's facial injuries and his media silence after the crash. Over the course of a month, one woman after another—stunning blondes, porn stars—stepped forward to claim that she, too, had shared wild nights with Woods. At least nineteen women confessed to affairs ranging from brief encounters in parking lots to "sext" messaging, three-ways, and more.

Tiger Woods was the top-paid athlete in the world, a billionaire, and was married to Elin Nordegren, a gorgeous young blonde Swedish model. Woods was universally respected as a role model for children. The Tiger Woods Foundation, which started out as a golf camp for underprivileged youth in 1996, now includes university scholarships; an association with Target House at St. Jude Hospital in Memphis, Tennessee; Tiger Woods Learning Center (a thirty-five-thousand-square-foot learning facility in Anaheim, California); and the Start Something character development program, which reached over a million young people in 2003.

Having women such as Veronica Siwik-Daniels, the star of *Big Breasted Nurses* and *MILFs in Action* step into the media spotlight to discuss the details of

Woods's sex life—including their unprotected sex, her alleged miscarriages, and abortions—torpedoed his status. As his wholesome, family-guy image collapsed, so did his sweetheart deals with huge corporations. AT&T, Gatorade, Gillette, TAG Heuer, and Accenture severed their advertising and endorsement dollars, costing him hundreds of millions in lost revenue. Even *Golf Digest* canceled his long-running series of golf instruction articles.

In February 2010, Woods appeared on television for the first time since the car crash that ruined his life. Appearing in a blue shirt and blazer before a small audience of reporters, he apologized profusely for his errors.

> *I am deeply sorry for my irresponsible and selfish behavior I engaged in. I know people want to find out how I could be so selfish and so foolish. People want to know how I could have done these things to my wife, Elin, and to my children.... I am also aware of the pain my behavior has caused to those of you in this room. I have let you down. I have let down my fans. For many of you, especially my friends, my behavior has been a personal disappointment. To those of you who work for me, I have let you down, personally and professionally. My behavior has caused considerable worry to my business partners.... I know I have severely disappointed all of you. I have made you question who I am and how I have done the things I did. I am embarrassed that I*

> *have put you in this position. For all that I have done, I am so sorry. I have a lot to atone for.*

The vision of contrition, Woods virtually begged the world to give him another chance. "I owe it to my family to become a better person, I owe it to those close to me to become a better man.... I have a lot of work to do, and I intend to dedicate myself to doing it," he said. "I ask you to find room in your heart to one day believe in me again."

Whatever you imagine Tiger Woods's private feelings were during that press conference, he publicly displayed guilt and shame about his behavior. This is what society demands. The parade of *mea culpas* from actors, politicians, athletes, and other public figures on cable news seems endless, as does the public appetite for it. News of revolutions in foreign countries is routinely interrupted by a breaking update featuring Lindsay Lohan saying she's sorry, again. Why do we care? Because we are inherently social animals, with large chunks of our mental machinery specifically designed by evolution to care. Not only can we not help it, our hunter-gatherer biology *encourages* it. We need to know we can rely on our tribe members, and when they mess up we want to make sure they won't let us down again. Natural selection has groomed us to pay very close attention to the goings-on in our group. Gossip, innuendo, who's-doing-what-with-whom: it was all vital survival information to our prehistoric ancestors. An unreliable member of the clan shirking duty could mean death to everyone. Trust and survival were completely intertwined.

Energy frugality is behind a lot of social emotion, particularly expressions of regret, guilt, shame, and embarrassment. In the small hunter-gatherer bands of our distant past, transgressions were a deadly serious business. People would kill each other over cheating, lying, stealing, and especially infidelity. Ms. Nordegren whacking Tiger in the head with a club has a long evolutionary history behind it, as did astronaut Lisa Nowak's nine-hour diapered road trip to kidnap a romantic rival. Yet all this violence is bad for the group. It kills cooperation and reduces the ranks. By allowing a transgressor to demonstrate publicly his understanding that this behavior is not acceptable, group cohesion can be maintained without a lot of wasted energy. Rather than losing a valuable member through fighting, banishment, or abandonment, the member's transgression can be acknowledged and punished. The group stays together and energy is conserved.

With regret comes guilt, the purpose of which is to keep us from hurting or neglecting others in our group. Biologists believe that guilt evolved from the instinct of animal mothers to care for their infants. It is a small leap from caring for children to caring for all members of the pack.

Shame, unlike guilt, evolved from an animal's need to protect itself from the aggression of other group members. Social animals have behavior systems such as social anxiety, flight, submission, and appeasement to increase attentiveness to others in the group who might inflict harm upon them. A display of submission might influence a dominant animal not to attack a subordinate. Human shame is a more complex version of these behaviors. Shame is not only about protecting the individual; it also serves to keep the group functioning properly as a whole.

Although we can feel such emotions privately, which presumably acts as a negative reinforcement for bad behavior, the main purpose of such emotions is for the benefit of the whole group. Researchers Gun Semin and Anthony Manstead created an experiment demonstrating the social purpose and power of embarrassment: They had actors knock over a sales display in a supermarket. They then asked subjects watching this mishap to rate the perpetrator in terms of likability. If the actor showed signs of embarrassment, the subjects rated him as highly likable. If the actor did not seem embarrassed, it didn't matter what else he did to make things better. Even if he carefully took the time to rebuild the sales display, he was deemed less likable than if he didn't help in any way and just showed embarrassment.

In situations like this, we can relate to the other's embarrassment at the mistake; we can feel what they feel because we've been there. We feel empathy. But there's a possible physical dimension to empathy that could be generating those feelings as well.

Researchers at the University of Parma in Italy placed electrodes in the frontal lobes of a macaque monkey, which would fire whenever the monkey grabbed a peanut. The scientists very reasonably decided that this neuron was a motor neuron, responsible for guiding the motion of the hand. However, one day the macaque was not moving at all when a hungry scientist walked into the lab, picked up a peanut, and ate it. The monkey's neuron fired, just as if it had picked up the peanut and eaten it itself. The team discovered that approximately 10 percent of the macaque brain contains these neurons, called mirror neurons, that cannot tell the difference between doing something and watching somebody else do the same thing.

Watch a group of fans during a football game. They're not just watching the action; their muscles tense, they jump, they wince and grimace. Their neurons are firing as if they're actually playing the game. The idea of "monkey see, monkey do" neurons has huge implications for learning and social interactions in primates. It has not been conclusively proven that mirror neurons exist in humans, but there is laboratory evidence to support the idea.

UCLA psychiatrist Marco Iacoboni believes that mirror neurons also help us tune in to the emotions other people are feeling. When we see somebody smile—which, after all, is a sequence of motor movements in the face—our mirror neurons light up and we feel a little emotional echo of that smile within us. Empathy, attunement, understanding the intentions of others: mirror neurons may be at the core of our social interactions, allowing us to mesh intimately with others. As Iacoboni says, "What do we do when we interact? We use our body to communicate our intentions and our feelings. The gestures, facial expressions, body postures we make are social signals, ways of communicating with one another. Mirror neurons are the only brain cells we know of that seem specialized to code the actions of other people and also our own actions. They are obviously essential brain cells for social interactions. Without them, we would likely be blind to the actions, intentions and emotions of other people."

V.S. Ramachandran, director of the Center for Brain and Cognition at the University of California San Diego, asserts that the discovery of mirror neurons "will do for psychology what DNA did for biology: they will provide a unifying framework and help explain a host of mental abilities that have hitherto remained mysterious and inaccessible to experiments." These abilities include language, theory of mind, and advanced toolmaking.

If it turns out that the human brain is rich in mirror neurons, they could be essential to the experience of social emotions and motivating us to behave empathically and compassionately. And it is more evidence that something other than the ego is in charge. No one consciously intends to mirror others' respiration and blood pressure. Their response is driven by the organism, beneath awareness.

***

It may seem as if social emotions are only good for the group, but they are also so important to individuals that losing them can be life threatening. Studies on brain-damaged individuals who have lost the ability to feel social emotions show a drastic diminishment in the ability to make good decisions. Previous to their brain injury, they were typically hardworking, successful individuals with families. After their injury, they show no impairment in intelligence or logic. They are able to reason their way correctly through tests designed to measure social skills. Even with all their reasoning skills intact, however, they often lose their marriage, experience strain in family relationships, get fired from their job, lose their independence, and cannot be trusted to make any decisions for their own well-being, especially financial ones. They lose their drive and ambition, and while they have logical skill, their lack of emotion makes it impossible for them to prioritize facts into commonsense choices—say, that getting up and getting to work on time might be a good idea in terms of keeping their job. The lack of shame or embarrassment means they no longer care what others think. Social emotions can no longer motivate and inform their behavior.

That the loss of social emotions such as guilt, regret, and shame led to reduction of competency was so counterintuitive that for many years researchers attempted to find other areas of such patients' brains in which undetected damage might prove responsible for the bad decisions. They looked for a damaged ability to recall useful details of their situation, and they looked for damage to their memory. All such investigation came up empty.

We can't live without social emotions. Yet despite their centrality to our lives, or perhaps because of it, they can become a trap.

***

Ground Zero, the ruin of the World Trade Center, was closed as a recovery site on May 30, 2002, more than eight months after the terrorist attack. As the officers of the Port Authority (PA) describe it, that day was "September twelfth" because it was the first time since 9/11 that they had taken a rest. They had spent every day of the intervening eight months in the "Pit," digging out bodies twelve hours a day, seven days a week, with few breaks and no rotation.

Officer Karl Olszewski described the work:

> *When you are in the hole and someone passes you the torso of what used to be your fellow officer who you knew and then they pass the skull of a woman he tried to save, that is when the gravity of the situation starts to hit you. But you*

can't stop because someone is yelling at you to pass another item. Then a container with body matter or another part comes by, and you see 20 red bags with bodies and other bags with body parts. Then you realize you are covered with remains and you smell of death. You wipe your hand near your mouth and you taste death and you reach for water and there is none. So you spit and spit. But you can't get the taste out of your mouth. You wash out your mouth and it's still there. You begin to think about it, and you finally walk out to get some fresh air. But you find yourself just waiting in line to get back into the pit because that is your job.

The Pit was seven floors deep and fifteen stories high, a tangle of bent steel and concrete. When body parts were found, Olszewski would dig very carefully around them, trying to keep the pieces intact and undamaged so a medical examiner would have a better chance of identifying the remains. "No matter where you turned, the same carnage existed," he said. It was "a gigantic pile, a pile of death." During those months he felt sorry and depressed, and broke down into tears.

Eight months of such work had left the PA officers in a state of trauma few have ever seen. Even the trauma experts sent to help them deal with their emotions said that the PA's situation was unprecedented. The grief of loss, the stress of the work under emergency conditions, the sadness and anger at the attacks themselves: all had continued to build up over time. There had been no time to process or release the pain. Officer Peter Hernandez, who patrolled the Towers daily before the attacks, said, "I knew a lot of civilians in there. I saw their faces in the morning, during lunch, coming to work, leaving from work. I knew a lot of them by name."

But there was a second set of feelings as well. The Port Authority was almost unknown in the media. "We felt ignored. It affected us department wide. We communicated with all the agencies; we were the first in and the last out. We were forgotten about. It is frustrating and it does hurt," said Hernandez. The NYPD and FDNY were celebrated as national heroes in the media, basking in the public outpouring of adulation and support. While the PA officers were stuck in a dusty hole, passing the body parts of their friends to the surface, some members of these more famous units were featured on magazines, television, and the Internet, as well as receiving perqs, parades, and accolades. Sergeant Michael Florie said, "A lot of companies were giving Disney vacations and such. Our guys had to stand on patrol and watch the other departments go on these vacations."

The Port Authority lost thirty-seven fellow officers in the 9/11 attacks, more than 2 percent of their force: the worst tragedy to befall a law enforcement agency in history. The officers attempted to cope individually with their overwhelming emotions of sorrow, grief, rage, and frustration. Officer Christopher

Roughan gathered broken hunks of white marble from the wreckage and used a rock saw to cut it into crosses, which he gave to the family members of deceased officers. "We all handle this in our own way," said Roughan. Frank Giaramita, a sergeant in the Port Authority, said, "It hasn't ended for any of us."

Karl Olszewski dealt with his feelings in his own way too. A member of the army reserve, he went to join the fight in Afghanistan. Still there in 2006, he wrote on a website:

> *Having covered nearly all of (Afghanistan), I earnestly seek to engage and destroy "our enemies." I was there on 9-11 and for the next nine months; I worked on both the rescue and recovery teams at Ground Zero. It is, for the most part, why I'm here.... I can never talk or write about this without tears in my eyes, so please forgive me for any spelling errors.*

The Port Authority officers gave as much as any human could for their society, short of sacrificing their lives. Despite doing what their altruistic emotions motivated them to do, some of them ended up feeling bitter and unseen. Denied their rightful place in the sun, they felt ignored, frustrated, and hurt. They had done their part, but society had not held up its end of the bargain.

This is the hidden downside of social emotions. It is the very success of social emotions and their centrality in our lives that turns them into a prison. Often we think of ourselves as individuals, as separate and alone. But we are intensely social beings, and most of our feelings, as well as most of our higher brain functions, evolved to facilitate social interactions. Instinctively we know that we need the group to succeed, and in turn, the group needs individuals to flourish. Our prehistoric drive to belong and to be seen as a vital, contributing member of the group occupies our thoughts day and night. It pushes us to the point of obsession. But now we know: the prehistoric, pleasure-seeking-nervous-system need to achieve is an aspect of ego function.

We are extremely sensitive to any signs that we are ignored, dismissed, or left out. Rejection can trigger awful, terrifying feelings, while signs of acceptance fill us with joy and happiness. Positive feelings of belonging are so intense that we will do almost anything to get them. Whether conscious of it or not, we spend most of our time thinking about how to be seen as a valuable member, an *asset to the group.* We want to look good. We try to hide our weaknesses and promote our strengths to secure our status.

Notice how the Port Authority officers were sensitive to the fact that they were doing a good thing—something society sees as helpful to the group—that they were not being recognized for. It was not enough to do the thing; there was also the natural desire to be acknowledged for doing it. This is what every human desires as a social being because positive reinforcement from the group insures you'll repeat the behavior that helps the group survive. Our genes mold our psychology.

As much as our desire to do good has motivated us in useful and even beautiful ways, it also is a major structural element in the prison of ego that confines us. Imagine what drove Karl Olszewski to go from Ground Zero to Afghanistan and risk his life daily. Was it pure altruism, or was there also an unfulfilled need to be seen as a hero? Instinctively we all want to be acknowledged and accepted, and we promote ourselves as valuable at every opportunity. We love compliments, flattery, and awards. We love to be loved, admired, and adored. We crave promotions and being singled out for praise at work. In school, we delight in getting A's, in being the homecoming queen or the most valuable player. These events signal that we've succeeded, we're safe and protected. We belong. We won't be left to die alone on the savanna.

Feeling like an asset to the group generates pride, which is the counterbalance to shame and embarrassment. Pride signals our dominance and status, and tells us that we are valuable. It's about the biggest ego boost there is. Pride, like shame, can be seen as having evolved directly from its animal precursors, which include dominance displays, often involving special postures and gaits.

The display structures of pride and shame, bodily gestures and physical postures, are deeply hardwired into the human nervous system. Scientists recently studied high-resolution, high-speed photography of the 2004 Olympics and Paralympics judo competition involving blind contestants. Across the board, when these athletes, who had never seen another person in their lives, won a match, they displayed the typical primate body language of pride—open posture and a puffed-out chest. Conversely, if they lost a match, again they exhibited the stereotypical primate shame posture: lowering the head, slumping the shoulders, and narrowing the chest. Not surprisingly, these physical expressions are the same across cultures.

The drive to be an asset is a remnant of our hunter-gatherer past and may be somewhat out of place in a modern context. Gone are the days when the slightest deviation from giving all for the group may have spelled disaster. In many parts of the world, people are not on the edge of starvation or in life-threatening danger from rival groups as everyone was a hundred thousand years ago. There is more than enough surplus energy for us all, even though we are not distributing it equitably. Just like the desire for sugar or for big, open lawns, our relentless need for approval from our group is mainly an evolutionary artifact. Yes, it helps us get along with others and aids in the smooth functioning of society, but the constancy and sharpness with which we feel it is unnecessary today. Instead, the terror of not belonging, the constant need to be seen as a hero or the best at something, gnaws away at us. It causes an endless game of comparison with other people that is a cause of real suffering. This is social emotions gone awry, paradoxically cutting us off from our more fundamental connections with other people and instead thrusting us into a never-ending, isolating game of one-upmanship.

Just like the basic emotions—happiness, sadness, disgust, anger, and fear—social emotions helped the human family thrive. But the time and energy we spend attempting to gain praise and acceptance in a world not of thirty people, but of 7 billion, is out of place and probably futile. Yet we are relentlessly driven to it by our genetic inheritance. Social emotions, in fact, compel us to exhibit some of humanity's least attractive behaviors. Ever been lied to by someone on a dating site? They are trying to make themselves appear to be a more desirable mate. Otherwise calm, rational scientists will fight like alley cats over who gets credit for a discovery. Backstabbing, lying, conspicuous consumption, cheating, double crossing, being twofaced: our vocabulary reveals how much we dislike false jockeying for status. We use character assassination and *ad hominem* attacks to raise our status by lowering the status of others. Calling people freeloaders, lazy, liars, or cheaters are some of our worst insults, because they bring into question someone's value to the group.

***

A human being is a part of a whole, called by us "universe," a part limited in time and space.

> *He experiences himself, his thoughts and feelings as something separated from the rest—a kind of optical delusion of his consciousness. This delusion is a kind of prison for us, restricting us to our personal desires and to affection for a few persons nearest to us.*

From the largest perspective, we are nought but animated dust, a paltry handful of chemicals in a puddle of water, stirred by a billion years of natural selection. With the rise of DNA and multicellular organisms, some cells began to specialize as sensors, becoming neurons. The evolution of the nervous system, the spinal cord, and the brain made it possible for animals to move and behave with ever greater intelligence. As the pleasure/pain alphabet of the animal nervous system grew more sophisticated, it added hormonal gear shifts, powerful drives, and physical expressions—emotions—giving it tremendously nuanced interactive possibilities with the environment, with others, with energy sources. Yet this rich vocabulary of behavior was bought at the cost of a biological bondage that is constantly poking and prodding, forever waving its carrot and stick.

Our emotional system kept our ancestors alive. It still looks out for us and enfolds us in a mechanistic push-pull relationship with our own experience. It makes us hug our children, makes us work all day to get ahead, makes us crash airplanes into buildings, makes us terrified and enraged, makes us rescue people from burning buildings. Einstein's prison is alive and well in our brain structure, our genes, and our lives.

This, then, is the dark side of our emotional inheritance. We are trapped inside a biological system that controls and manipulates us constantly. Trying to feel good and avoid feeling bad occupies our waking hours. Many of the things we think of as most valuable, personal, and intimate in our emotional lives are genetic tools designed to domesticate us. Even the person we believe ourselves to be may turn out to be largely a contrivance of natural selection.

PART TWO

The Prison of Thoughts

chapter 6

becoming human

In 2002 Professor David Klett wrestled with his memories of what he had taught a young Muslim student in the 1980s. "We covered ... the fundamentals of jet engines and propulsion and chemical reactions, combustion reaction ... and those things would have been necessary for them to at least consider when they planned the World Trade Center attack with the airplanes," he said.

Klett had good reason to lament the lessons; the young man he taught was named Khalid Sheikh Mohammed, the "Forrest Gump of al-Qaeda," perhaps the most effective terrorist in history. Khalid Sheikh Mohammed has been implicated in the 1993 World Trade Center bombings, the Bali nightclub bombings, the failed "shoe bomber" plot, the Millennium Plot, and the murder of Daniel Pearl. But his main claim to fame is that he was the man who dreamed up the September 11 attacks.

Khalid Sheikh Mohammed went from his graduation in North Carolina to Afghanistan, where he and his brothers fought in the anti-Soviet jihad. He served as a secretary to mujahedin commander Rasul Sayyaf and briefly met Sayyaf 's friend and collaborator Osama bin Laden. Sayyaf and bin Laden had put together a training camp for fighters, and Khalid

Sheikh Mohammed's nephew Ramzi Yousef was a graduate. Ramzi Yousef had used his training to detonate an enormous truck bomb in the basement of the World Trade Center in 1993. This failed to knock over both the Twin Towers, the goal of the terrorists, but the attention and respect Yousef gained was not lost on Mohammed.

Khalid Sheikh Mohammed met Yousef in the Philippines in 1994, where the two hatched a number of terror plots. They imagined something really big, a super-terror plot called "Bojinka" (meaning "boom" in Serbian, a word Khalid Sheikh Mohammed had picked up in Afghanistan) to blow up twelve American airliners over the ocean in the hopes of bringing all air travel to a halt. This plan would eventually mutate into the 9/11 attacks.

Yousef was a master bomb maker who proceeded with a series of cold, step-by-step experiments. He created an undetectable nitroglycerin bomb for Bojinka that he tested in a Manila movie theater because the seats were similar to airliner seats. The bomb blew up, injuring several people.

Encouraged, Yousef made a more advanced version of the device, one-tenth the power planned for the Bojinka bombs, and smuggled it himself onto a flight from Manila to Tokyo. Using a contact lens fluid bottle to hold the clear liquid nitroglycerin and a Casio watch as a timer, he deposited the bomb under a seat on the jet and then deplaned. Two hours later it detonated, killing a Japanese man, blowing a hole in the cabin floor, and nearly bringing the plane down.

Pleased with the second test run of his device, Yousef began assembling the dozen large bombs to be used in the Bojinka plot. Then things went awry. He accidentally started a chemical fire in his apartment and was forced to flee. Besides all the bomb ingredients, police also found and decrypted a computer hard drive left behind, discovered his plans, and alerted the FBI.

Yousef was later arrested in Pakistan, but Khalid Sheikh Mohammed was still enamored of Operation Bojinka, the details of which had changed in the meantime. He met with bin Laden in 1996 and presented his new plan to crash airliners into buildings in America. Bin Laden did not OK the plan, but the idea for what was known as "the planes operation" was now in the air.

Over the next couple of years, Khalid Sheikh Mohammed's vision of the planes operation continued churning in his imagination. It now called for a total of ten planes to be hijacked, five in America and five in Asia. The American planes would be crashed into nuclear power plants as well as FBI and CIA headquarters. Mohammed wanted to hijack one plane himself as a special political gesture. He would kill all the men on board, make a pronouncement condemning U.S. policy on Israel, then land the plane and release all the women and children.

In late 1998, bin Laden called Khalid Sheikh Mohammed to an al-Qaeda training camp near Kandahar. Al-Qaeda had meanwhile been toying with its own version of the planes operation. Bin Laden believed that if he could strike the centers of American government and power, the US would be not only be crippled as a superpower, but the federal power over the individual states would dissolve and America would collapse completely. He said that American power "...is built upon an unstable foundation which can be targeted, with special attention to its obvious weak spots. If it is hit in one hundredth of those spots, God willing, it will stumble, wither away, and relinquish world power."

Together with Khalid Sheikh Mohammed, he now set about planning which targets to hit in order to bring about his vision. They imagined what would do the most damage to American power. Mohammed wanted to complete his nephew Yousef 's dream of bringing down the World Trade Center. In all they planned for ten aircraft to be hijacked from both the East and West Coasts of America, which would be crashed into buildings, including the White House, the Capitol, the Pentagon, Los Angeles's Library Tower, the Sears Tower in Chicago, the Space Needle and the Bank of America Tower in Seattle, the Empire State Building, and the World Trade Center.

The plan required men who could pilot commercial airliners. Bin Laden assigned his most trusted agents, including Khalid al-Mihdhar and Nawaf al-Hazmi. They were highly trained, fearless, and ready to martyr themselves but had no experience with aircraft and even less with Western culture.

Just at that moment, Mohamed Atta and other members of the Hamburg terror cell arrived for terror training in Kandahar. Highly educated in technical fields, used to living in the West, and some fluent in English, they were immediately singled out. Over a special Ramadan feast, bin Laden revealed that they had been chosen as principal agents in the planes operation. Atta would be the leader, taking his three men to America to undergo professional training as commercial airline pilots. After almost a decade of planning, Khalid Sheikh Mohammed's vision of using aircraft in a massive terror attack was at last becoming a reality.

Manifesting the 9/11 attacks required an intricate dance between imagination, memory, and emotions. Driven by the anger he felt about America's policy toward Israel and its effects on Muslims, Khalid Sheikh Mohammed allowed emotion to direct his behavior and manifest in action. He could learn and remember the tools of the terrorist trade, like how to hijack planes and which symbols might be important to attack. He could imagine how to rearrange the items in his memory to create a new outcome in the future. He had at his disposal additional capacities far beyond those of any animal: the power of the human brain to imagine the future and make a plan to shape that future. Instead of acting on his emotions instantly, like an animal would, Khalid Sheikh Mohammed was able to wait years for his motivation to lead to action.

***

> *A human being is a part of a whole, called by us "universe," a part limited in time and space. He experiences himself, his thoughts and feelings as something separated from the rest—a kind of optical delusion of his consciousness. This delusion is a kind of prison for us ...*

We have used the miraculous supercomputer in our heads to overcome disease, reduce poverty, understand the world we live in, create labor-saving machinery, and become the most successful creature on the planet. On the other hand, we have also used it to design death factories, plan genocides, create nuclear weapons, and ravage the planet.

Khalid Sheikh Mohammed's hatred twisted his imagination to create a terrible vision of death and destruction. We have captured Khalid Sheikh Mohammed and put him in the prison at Guantanamo Bay; his nephew Ramzi Yousef resides in the federal Supermax prison in Florence, Colorado, alongside American terrorist Terry Nichols, shoe bomber Richard Reid, Unibomber Ted Kaczinski, and 9/11 conspirator Zacarias Moussaoui. Their crazy, violent, emotionally disturbed minds are locked away from society.

But any one of us can experience a lesser version of the same wild internal cocktail of thoughts and feelings. Ideas come out of nowhere, triggering strong emotions, which foster more thinking, which results in more emotions. Animals only react to their immediate physical environment. If you are a mother jackal with pups, caring for them is your immediate and total concern. Humans are different. We react to our *mental* environment, the thoughts in our heads, with the same intensity as if they were real. The trouble is, we misunderstand and misapprehend, yet our emotions will be triggered—our motivation and guidance engaged—as though we are defending an indisputable truth. This is part of the prison of thoughts and feelings that Einstein is referring to, one that can cause so much misery and be so difficult for us to endure.

No other animal has this problem. Even chimpanzees, with their relatively high intelligence, react emotionally to their external circumstances rather than to the phantasms generated in their brains. Humans live in a biologically generated virtual world constructed moment by moment in our brains, based on memories and fantasies, with input from the world around us making up only a fraction of what we respond to as reality. We react to our imagination, lost in mental worlds that often bear little relationship to the reality around us. Our emotionally charged mental constructs are the result of millennia of natural selection. They had a purpose.

***

Squirrel monkeys love to eat dates. If you offer them a choice between one piece of date and four pieces, they will always pick four pieces. But if you link the dates with something else, the monkeys' behavior changes. If, for example, you take away their water for three hours every time they pick four pieces of date, the monkeys will start choosing one piece instead of four. Is this an example of planning in animals? Do the monkeys anticipate how long they will have to wait to have a drink of water if they pick four dates, and so overcome their desire and choose only one? Can they imagine the future?

Nervous systems evolved as a way to guide organisms toward new sources of energy and away from sources of damage, or energy loss. When neurons gather into brainlike processors, the ability of the nervous system to predict successful behavior grows. Evolution equipped animals with a rote set of behaviors and behavioral cues that were roughly the same in all individuals of a species. They also had a memory that allowed individuals to augment their instinctual behavior with *learned* behavior.

For example, all rabbits have a programmed fear of coyotes, but an individual rabbit might remember that the last time it was near a certain watering hole it was attacked by coyotes. The rabbit has learned to avoid that place, and this memory will help it survive. So its general fear of coyotes gets attached to the memory of the danger of meeting them at a specific watering hole. Virtually all animals exhibit this kind of energy-conserving behavior in which a general instinct (fear of predators) helps them learn specific individual behaviors (avoid the certain place where the predators are). This could be called the beginning of planning, because to plan we have to use our memory of places, other creatures, and so on to build a picture of the situation. But at this level it is rudimentary.

On Argentina's Valdes Peninsula the beach is packed with seals that have hauled out onto dry land in order to escape the hungry killer whales prowling the waves. The surprise is on the seal, however, because a small group of orcas there have learned to intentionally beach themselves just long enough to grab a seal in their massive jaws. Picture a twenty-five-foot-long killer whale slamming onto the shore, water flying, shaking a seal in its jaws like a rag doll.

This is not an instinctual behavior, and it is extremely dangerous for the orca because they can get stranded and die. It is not passed down genetically, it is learned; scientists have observed the older male orcas teaching the young ones. In the Crozet Islands off India, it is the adult female orcas who have learned this intentional beaching maneuver and teach it to the youngsters. A mother orca actually pushes her young onto the beach and waits nearby, ready to rescue it if it is having trouble getting back into the water. Scientists have even seen the skilled mothers teaching intentional beaching to the young of other, less skilled mothers.

So although orcas have an instinctual capacity to hunt seals, these individuals have learned to get seals in a way that was not preprogrammed by evolution. They have enough intelligence to see a seal on the beach and *plan a way to get it.* The killer whales can even remember that this behavior is quite dangerous and predict that young orcas will need to be trained to do it safely.

Orcas are the top predator of the sea, even capable of hunting great white sharks, and are quite intelligent. But we have yet to see an orca executing a behavior that had to be planned days or even years in advance, like a human being does. A mother planning her child's birthday dinner a month in advance is a feat of future imagination that is orders of magnitude too difficult for an orca to manage. Creatures other than human beings just don't have enough brainpower to pull off long-range prediction.

The average intelligence of human beings worldwide is more or less the same. We can all speak, make art, create culture, plan into the future, and remember our past. The archaeological and genetic evidence points to the fact that the conceptual revolution took place all across the world at almost the same time. Once one human somewhere accidentally evolved a smarter brain, it proved to be such an energy advantage that it quickly spread to all human beings everywhere in just a few thousand years.

This increase in intelligence allowed us to *imagine.* The evolutionary purpose of the imagination is to plan and predict. Planning and prediction rely on memory: what happened before is the predictor of what will probably happen again. In us, the nervous system reaches its highest peak: imagination, planning, memory, and emotion, all working together. Yet it took a long time for evolution to get us to where we are today.

It was about four million years ago, on the new African savanna, that we first did something we now see as a milestone in the life of every child: we went from crawling on all fours to standing up. *Australopithecus*—the ancestor everyone knows as "Lucy"—was the first proto-human to have a skeleton adapted to continuous standing. She had a brain only one-third the size of ours and just a bit larger than a chimp's. Yet her bipedalism would drive her kind toward having not only bigger brains, but developing the brain's best tool: *hands.*

In most animals, the forelimbs are just a second pair of legs. Even chimpanzees mainly use their arms either to hang from trees or as front legs for walking. But when we stood up, our front legs were set free to serve entirely new purposes. It allowed our hands to develop into refined instruments capable of delicate manipulation.

Four million years ago we hefted rocks to smash open nuts, like modern apes do, but we did not yet have the imagination to shape a rock into a tool.

Our diet was fruits, vegetables, and roots, as we could not yet cooperate enough to hunt as a team.

Australopith society was not about cooperation; it was a rigid hierarchy of alpha males like that of chimpanzees. The males were as much as 50 percent bigger than the females. In contrast, modern human males are on average only 15 percent bigger than females. Big differences in the size of males compared to females are the result of intense rivalry for mating partners. Because the male who won a fight got to spread his genes far and wide, evolution favored hulking brutes that could dominate the competition.

Australopiths were born, lived, and died on a grassy plain, mutely watching the days pass as we fed, raised our young, and kept the wolves at bay. There were no formal social structures, no language, no real tools, no culture whatsoever. Yet in our upright posture, our growing brain, and the increasing dexterity of our hands, the seeds of revolutionary brain change were already growing.

By roughly two million years ago, the *Australopiths* had evolved into *Homo habilis,* or "handy man," and had developed enough brainpower and manual dexterity to begin making tools. The feedback loop that would lead to imagination, poetry, space shuttles, and the Internet was now in full swing. Brain growth was driven by the ability to work with our hands; manipulating objects with our fingers directly increased our intelligence. Those who can use their hands more skillfully have an advantage, and bigger brains have that skill. So evolution selected for intelligence. And the better our hands got, the more food we could get. Using a sharp stone tool to butcher an animal and get at the nutrient-rich bone marrow gives you extra food resources (energy) and saves work (energy). You can pound tough roots to make them easy to digest. The smarter you are, the more food you can get, the smarter you can become. It's a positive feedback loop.

Tools accelerated evolution. With tools you get a massive boost in energy gathering and storing without having to wait for evolution to develop new parts of your body. Chimpanzees will sometimes strip a twig and use it to get termites to eat. Imagine what it would take to evolve a long, thin finger specially adapted to gather termites. It would certainly be possible, but it might take tens of thousands of years. If, instead, the chimp brain becomes just smart enough to turn a twig into what is effectively a specialized finger, it can get the termites and benefit from the food energy today.

With a big brain, you can fashion what are essentially new body parts. Google functions as a giant memory bank; clothes are a sort of protective fur; and a car is a really fast pair of legs. Natural selection strongly favors animals that make tools, although it is a very difficult thing for them to do. It takes a combination of lot of brainpower and a lot of dexterity. Whales may have the brainpower to create excellent tools, but we will never know because they don't have hands with which to make them. Elephants have high intelligence but an awkward trunk that can grasp only large things. In the case of humans, a unique and fortunate combination of climate, posture, brain size, environment, and social organization came together in a feedback loop that developed this toolmaking ability.

The craftsman's masterpiece, the cutting-edge high-tech stone tool invented by *habilis,* is known as a chopper. A chopper is a broken rock with a sharp edge. To make a chopper, you take two round river stones. Smack one of them on the edge with the other one. Pieces of the stone will chip off, leaving a sharper edge. If you are feeling especially creative, turn it over and knock some flakes off the other side as well. Now you have a tool that is smooth and round on one end, which makes it rest comfortably in your palm. The other end, where the business gets done, is a sharp, jagged edge.

It may not seem like much, but the mental steps required to craft a chopper made *Homo habilis* a supergenius compared to a chimp. An ape can use a rock to break open a nut, just as it can sit in the sun to get warm. But to look at a rock, envision what it could *become* and picture the steps necessary to manifest that vision was a brand new skill. To look at a rock and see a tool in your mind is the same as to look at an airliner and see a guided missile. The difference is just one of scale, emotional influence, and intention.

The chopper fueled our success. It is a wonder of the imagination, a kind of Ur–Swiss Army Knife with an axe, hammer, knife, and awl rolled into one. Choppers were just the energy savers we needed to gain a competitive edge over larger predators. And this success led to more evolution in the same direction: more brains, better tools, more complex social organization, and a revolution of the imagination.

***

"When the plane struck the building it felt exactly like an earthquake," said Susan Frederick. She and her colleagues, who were on the eightieth floor, had no idea if it was a bomb or what, but they evacuated quickly. The elevators had been taken out by the impact, and so they made their way through the maze of various stairwells. Some were blocked, and Frederick had to find other ways that were open.

> *We finally got out of the smoke when we hit the 35th floor. It felt great to breathe fresh air and lifted everyone's spirits. We also started running into building personnel. Around the 27th floor we ran into firefighters climbing up. I can't imagine what it must have been like to walk up that many flights with all the gear they had. They looked so winded at that point. I doubt that they made it out before the building collapsed and my prayers and thoughts are with them and their families now.*

Frederick made it out safely, a scant four minutes before Tower 1 collapsed, because of the bravery of the firefighters who put themselves in harm's way to help. As Frederick guessed, the members of FDNY did not make it out alive. The world watched in shock as Tower 2 collapsed first, but the people inside Tower 1 did not know. Trapped in dark stairwells, they only felt another giant rumble.

Meanwhile, police helicopters buzzed close to the top of Tower 1, looking to see if it, too, was in danger of falling. "About fifteen floors down from the top, it looks like it's glowing red," said one pilot. "It's inevitable." Seconds later, another pilot radioed: "I don't think this has too much longer to go. I would evacuate all people within the area of that second building." The police on the ground heard these warnings, and most in and around Tower 1 were able to escape in time.

The firefighters in the tower didn't receive the orders from their own department to get out. Their radios failed repeatedly, and even if they had worked, they were not linked to the police broadcasts that would have informed them of the evacuation order. At least 121 firefighters who could have gotten out in time were killed when the tower collapsed, due to lack of communication.

Modern human brains are sophisticated planning machines. Unlike all other creatures, we can look into the future, predict possible scenarios, and formulate procedures to deal with them. There was supposed to be a plan in place for a catastrophe like the one on September 11. The fire department's radio system had failed in the WTC when the 1993 bomb exploded in the parking garage. As a result, Mayor Guiliani created the Office of Emergency Management in 1996. Twenty-five million dollars was spent attempting to coordinate emergency response measures, but no exercise drill had ever been done at the WTC that included the police, fire, and Port Authority departments together. Members of the FDNY and NYPD barely communicated at all during the 9/11 attacks, neither coordinating strategy nor sharing vital information. The plan failed.

When fifty fire officials in December 2001 took part in a planning exercise at the U.S. Naval War College, the college evaluators concluded: "As a function of command and control, it was evident that the Fire Department has no formal system to evaluate problems or develop plans for multiple complex events. It was equally evident that the Fire Department has conducted very little formal planning at the operational level."

In a way it's hard to fault the fire department. They demonstrated bravery above and beyond anybody's expectation. Planning operations for one department in a modern megacity must be difficult, let alone coordinating them with other departments and practicing them in various locations. By the time Susan Frederick got out of Tower 1, things were in chaos.

By now we were wet and covered in this ash. People all looked like their hair had turned prematurely gray. We were told to walk quickly up the street. Within minutes (we now know it was no more than 4) we heard a rumble, turned to see our tower begin to collapse and a large cloud of black moving up the street. We ran.

It is by God's miracle alone that I am convinced I got out. I am grateful to be alive and grateful for my family and friends. Amazingly, I never felt afraid and I believe that was because I truly felt God's hand upon me. It was not my Time. I'm not sure what is next. But for now smelling the flowers is just fine with me.

We take it for granted that we can think about next year's vacation, anticipate price increases by booking our travel and hotel early, determine location by looking at weather patterns, figure out the best time to take off from work, making sure nothing falls through the cracks while we're gone. We easily manipulate physical objects with our hands and *symbolic* ones in our minds; we strive to coordinate intricate, interlocking minutiae daily. And on a societal level, the organization of our efforts is what it's all about. The U.S. government, for example, employs almost 2 million people in about thirty different departments and agencies. Our ancient ancestors could not even dream of planning on such a grand scale, let alone accomplish it.

Social cooperation toward a common goal was minimal for *Australopithecus* and had grown a little more complex in *habilis,* with their sharing of work around a common fire, butchering their kill with their choppers. And as our brains grew more complex, so did our social interactions.

About 1.8 million years ago, a new and much improved member of our family tree came onto the scene, *Homo erectus. Erectus* was much smarter, with a brain capacity 75 to 95 percent of a modern human's. As the dome of the skull swelled to contain that intelligent brain, the size of the jaw shrank. We had almost completely lost the large, powerful teeth and jaws of an ape and were beginning to look almost human. We were taller too, standing at an average of about five feet ten inches—larger than we are today. The size difference between males and females had shrunk significantly, from 50 to 25 percent, evidence of our increasing social equality.

We may even have been able to communicate with each other beyond physical expressions of emotion. Although language had not yet been invented, fossils indicate that we had developed the larynx and Broca's area (the language center in the brain) required for rudimentary speech. *Erectus's* larynx was in a position similar to that of a modern two-year-old, and our ability to speak was probably on the level of a toddler. These barely articulate noises made a huge difference in our level of social interaction. We could communicate, organize, and cooperate much more effectively, and—as usual—this meant that we could save energy.

It was very hard to hunt alone, but a group of people could work together, like a pack of wolves, to surround, trap, and bring down large animals. Instead of competing constantly for rank and mating, the men could now coordinate as a team. Working together was a new thing in our history, representing a change of lifestyle. It was as if we switched from an individual sport like track to a team sport like basketball. Natural selection strongly favored this development because it gave humans access to the much more concentrated caloric energy source of fresh meat. Hunting gives you more calories because you can get the choice bits of animals rather than cleaning up leftovers. Our diminutive ancestors the *Australopiths* had been prey for large predators. Now we had become the predator, able to hold our own against the other animals in the savanna.

For group hunting to work, the social structure had to change from a rigid hierarchy to a connected, egalitarian system. If one alpha male had all the females as mates, that would have destroyed the cooperation needed for hunting. Thus began the system of pair-bonding: one partner for each individual. This new social organization, and new style of energy gathering, became what we know as the hunter-gatherer lifestyle, which was the way our species organized itself socially for almost the next

2 million years. Hunting in groups, gathering in groups, cooperating as a team, and working together for the good of the tribe formed our central organizing principle.

Our brains are still optimized for that ancient society, perfect for sitting around a campfire, picking greens in a meadow, or walking with our friends in the forest. If you enjoy working together with other people toward a common goal, it is because of the cooperative structure of *Homo erectus.* Whether foraging together for herbs, tubers, and fruit while helping each other care for the children—in the case of women—or coordinating a caribou hunt—for men—the deep structure of our brains is finely tuned to enjoy group effort. Our complex social emotions evolved at this time: our desire to be an asset to the group, our need to avoid shame and embarrassment. All of these evolved in order to keep these little hunter-gatherer groups working smoothly and functioning as a productive unit.

As natural selection began to favor proto-humans that could cooperate, the value of each individual began to grow. In a win-win society, you want everyone to do well. Each member of the team thriving helps the whole team flourish. If a member is sick or injured, that is one less person to help out. We began to evolve real empathy. Natural selection pushed us toward caring more about each other, something that we can find traces of in the fossil record.

Turkana Boy, an *erectus* fossil discovered by Richard Leakey's team, is our first physical evidence of the expression of empathy. He was probably about eight years old, which for *Homo erectus* would have been almost an adult. The fossil shows a severe abscess in several teeth that had eaten away a portion of his jawbone. This meant that Turkana Boy had to be cared for by others to survive as long as he did. Another *erectus* skull showed that an old man had lost all his teeth but managed to live several more years. His relatives would have had to mash food for him so that he could survive. The traces of compassion are there to read in the bones.

Homo erectus was on the verge of true humanity. The impulse-driven upright animal had slowly developed into something completely new: a being capable of formulating ideas and making tools, one that recognized the need to care for the ill instead of abandoning them, a creature that saw the benefit of cooperation. We created and tended fires, gathering around them to share warmth and community. We cooked food together and kept predators at bay by standing watch in shifts. We helped each other get food, took care of the children as a group, and learned to communicate our emotions to one another.

Erectus was wildly successful, covering the earth as far as we could walk—Africa, China, Indonesia, and Europe—adapting to every environment, and continuing to grow smarter and more capable.

***

Imagine for a moment a childhood home of yours. Now picture the same home with a neon-pink elephant inside. The elephant is playing patty-cake with a zebra-striped mouse as they both sing, "patty cake, patty cake, baker's man..." Really see and hear this in your mind. You have just done something no *Homo erectus* would be capable of: visualizing and bringing to life things that have never existed. Music, songs, completely articulate speech, houses, games: all of these things were still to come.

A new human species was emerging, *Homo sapiens,* who would be capable of formulating conceptual thought, using that capacity to make gathering energy resources for survival more efficient than ever. The capacity would develop into the ability to plan far into the future. And when thoughts unite with emotion to create belief, they can bring out our most elevated humanitarian instincts—sending firefighters inside the World Trade Center right into harm's way—or they can go terribly wrong as they did for Khalid Sheikh Mohammed and bin Laden. Powerful forces were coming together—imagination, concepts, emotion, group bonding, toolmaking—and their confluence would create an evolutionary perfect storm unlike any the world had ever seen.

chapter 7

the conceptual revolution

Just before the ninth anniversary of the 9/11 attacks, the Dove World Outreach Center, a small evangelical Christian church in Florida, announced "Burn a Koran Day." Pastor Terry Jones, author of the polemic "Islam Is of the Devil," invited Christians around the world to burn Qur'ans in remembrance of the victims of the 9/11 attacks. Despite the small size of Jones's congregation (about fifty), his pronouncement garnered round-the-clock press coverage and a firestorm of protest worldwide. "[Islam] is a violent and oppressive religion that is trying to mascarade *[sic]* itself as a religion of peace, seeking to deceive our society," said Jones.

Hearing of the plan, hundreds of Muslim students gathered in Kabul, Afghanistan, to protest the burning, chanting, "Death to America!" under the concerned eyes of U.S. troops. In Faizabad thousands of enraged Muslims demonstrated against the burning and attacked a NATO base. There were more protests and incidents in many other Afghan provinces, as well as in Pakistan, Indonesia, Somalia, and India.

The World Evangelical Alliance said that the event "...dishonors the memory of those who died in the 9/11 attacks and further perpetuates unacceptable violence." The Church of Jesus Christ of Latter-day

Saints, the International Humanist and Ethical Union, and many other organizations condemned the burning. U.N. Secretary-General Ban Ki-moon and President Obama condemned it, and twenty religious leaders in Gainesville, Florida, the home of Jones's church, called for local citizens to support Muslims "in a time when so much venom is directed toward them."

But Jones was not alone in his anti-Muslim feelings. On his program *The 700 Club,* television evangelist Pat Robertson stated that Islam "is not a religion" but "a violent political system bent on the overthrow of the governments of the world and world domination." Earlier he said that "If you get right down to it, Osama bin Laden is probably truer to Mohammed than some of the others" and also compared Mohammed to Hitler. Jerry Falwell expressed similar views, saying that "Mohammed was a terrorist" on *60 Minutes.* And evangelist Billy Graham's son Franklin, who, like his father, is a minister with a large following, stated that Islam is "a very evil and a very wicked religion," "a religion of hatred, a religion of war."

The Qur'an is considered to be the finest work of Arabic literature; Muslims believe it to be the Final Testament, the literal word of God, the ultimate holy book. Islamic scholars hold that the poetry style of the Qur'an cannot be recreated by human beings, and that the text contains miraculous prophecies unlike any other book. Like other holy texts, the Qur'an arose in response to questions of who we are, our place in the universe, how we should behave, and what might happen after we die.

Humans are the only animal to evolve conscious self-awareness, anxiety about the future, and a need for codes of behavior to aid in our ability to trust and cooperate with each other. When we do something that aids our survival, we get a pleasurable emotional punch from it, a physical reward that makes us want to do it again. If concepts such as those in the Qur'an become associated with survival, those ideas become doused with positive emotional feedback and voilà, we have conviction. Belief. If you surround yourself with others holding similar beliefs, planning and predicting become much easier propositions, since everybody agrees on the starting point. If we grew up in a culture that depends on the ideas in the Qur'an, then when someone agrees with us about their validity, we feel a powerful, positive emotional kick. We feel bonded and close to that person, safe and at ease. If another person questions the validity of these concepts, it's as if our life, our family, our society have been attacked. Rage and terror fill us, and we go into an acute defensive reaction.

Natural selection favored a mind that reacts excessively to itself, to its own thoughts, as strongly as if ideas were the external reality that first sparked them. As a result of the September 11 attacks, the minds of many confused the threat that came from a few dozen crazy people hijacking airplanes with a general threat coming from all Muslims. Instead of focusing on the few violent perpetrators who committed the acts, the mind goes immediately to *all Muslims are violent; all Muslims want to kill all Americans.*

It is this internal reality gone crazy that pushed a man to beat a Brooklyn cab driver named Osama to within an inch of his life. It is this impulse that pushed another man in Arizona to shoot a Sikh, thinking he was getting revenge against Muslims by killing a man in a turban at a gas station. It was this internal reality that Jones reacted to in calling for the burning of the Qur'an. And it is the attack on belief that caused the worldwide protests in response.

The association of Islam with the World Trade Center in flames, American soldiers being blown up, enraged foreigners with angry faces shouting in strange tongues, is unavoidable for the Stone Age brain we still carry with us today. The brain's wiring evolved to be dominated by fear instead of fact. More than 1.5 billion people are Muslim—almost one quarter of all living human beings. How many of these individuals are actually terrorists? How many are living normal peaceful human lives, going to work every day, hoping for a better life for their children? The group of nineteen dedicated terrorists who perpetrated the September 11 attacks is a vanishingly small percentage of all Muslims.

Natural selection favors fear. Organisms that are afraid will err on the side of caution and thrive, while intrepid others go too far and become another creature's lunch. This "better safe than sorry" feature has been wired into biology probably as long as there has been life. It is a conservative, protective urge, a bias toward safety, that is inherent in all nervous systems.

When a jellyfish, which has the most rudimentary nervous system, bumps into something, its whole body retracts, whether or not the contact is harmful. As nervous systems become more complex, the better-safe-than-sorry program of behaviors also grows more complex, yet fulfills the same primordial function: be afraid first; ask questions later.

The human nervous system continuously monitors the environment, hyperalert to danger. Our brain automatically runs better-safe-than-sorry worst-case scenarios based on what we perceive. If we ran "everything's coming up roses" scenarios first, we'd most likely trust the wrong people or head for the wrong watering holes, waving and smiling at the nice lions right before they turned us into lunch. A brain that is hardwired to imagine worst-case scenarios first and worry about bad things that haven't happened has what is termed *negativity bias.*

Negativitybias—awell-researchedphenomenon—focuses the majority of our attention on what can go wrong. It almost forces non-Muslims to think *terrorist* every time they see a Muslim even while the probability of a Muslim being a terrorist is roughly the same as you getting struck by lightning while reading this book. The nervous system does not care about probability: it sees the potential for danger and focuses on that to the exclusion of all else. Negativity bias doesn't automatically consign us to a life of pessimistic gloom, but it means that we often feel more worried than need be, and react to small concerns as though they were life-threatening.

When Albert Einstein wrote, "[A human being] experiences himself, his thoughts and feelings as something separated from the rest—a kind of optical delusion of his consciousness. This delusion is a kind of prison for us..." he was pointing toward the effects of negativity bias. Our thoughts, fantasies, and imagination trigger emotions, and negativity bias means that most of the time the emotion we will feel trapped by is fear or one of its sidekicks: anxiety, worry, stress, neurosis.

Things have changed since we evolved these reactions. In a hunter-gatherer's world, the range of things a person could worry about was relatively limited: eating or being eaten, maybe rain, snow, and cold. We got scared—the sympathetic nervous system ramped up heart rate, blood pressure, and stress hormones—and then it was over, and the parasympathetic nervous system had a chance to return the body to normal. Amplified by modern mass media, negativity bias catalyzes a witch's brew that our biology is ill equipped to handle. Bombarded hourly through media with concepts that make us afraid, the nervous system overloads. And when we add imagination to fear and push that button 24/7, we're pouring gasoline on a fire.

In 1994, there was a health scare about flesh-eating bacteria, which began when a British tabloid ran the headline "Killer Bug Ate My Face." American news media seized on the idea that these antibiotic-resistant superbugs were more powerful than anything seen by modern medicine and ran one terrifying story after another. In fact, this bacteria was nothing more than group A strep, something that had been well known for decades. Out of 20 to 30 million strep cases in the United States each year, between five hundred and fifteen hundred people suffer from the "flesh-eating" version, called necrotizing fasciitis. Your chance of contracting this form is tiny, but if you were alive in 1994, you were racing to the bathroom mirror to check for flesh-eating evidence on your own face.

While our brains evolved this protective stance, we didn't evolve the capacity to immediately apprehend the antidote: looking at the facts and understanding the math involved. Only a thousand people out of a population of 350 million get the disease each year. The possibility of it infecting us is almost nil, but the intensity of the news combined with the negativity bias and our imagination makes it seem urgent and dangerous. To the appearance of danger, our body reacts with very real, very understandable, and completely unnecessary fear.

If the news were limited to one such terrifying story on one day out of seven, the consequences would be minor. Instead, every day we are fire hosed with the seemingly imminent possibility of a nuclear terror attack, germ warfare, contracting a fatal disease, our children being attacked or killed, our spouse cheating on us or leaving us, our job drying up and blowing away, the country being overtaken by some extremist group or another, foreign economies destroying ours, earthquakes, fires, volcanoes, and thousands of other horrifying but remote possibilities. And as you read this list, you are probably thinking, *But I've seen all of this happen. How can you tell me not to be afraid?* We get caught in a vicious cycle of escalating fear that won't allow us the mental space to apprehend reality and take a deep sigh of relief.

Our ancient mammalian stress reaction was not built to withstand this level of concentrated, relentless fear. We react to these unlikely dangers with real terror, and our bodies flood with the damaging stress hormone cortisol as if we are under attack by many lions all day, every day. We end up in a state of chronic anxiety, pushed against a wall of stress we didn't evolve to withstand.

The conceptual abilities at the root of the trouble were very adaptive: they came about for one purpose, to give us the power to get more energy more easily. Imagination, prediction, planning, and language gave us the ability to expend less energy in gathering food and more energy trying to create an easier and more efficient life for ourselves.

Negativity bias isn't the only way the imagination runs wild. Our conceptual abilities allowed us to manifest complex visions our tree-dwelling ancestors could never have imagined. And the results were something nobody could have predicted.

***

High on a hilltop overlooking a grassy valley in southeastern Turkey is a complex of enormous stone columns sixteen feet tall, each weighing seven tons. The T-shaped columns are arranged in four circles about thirty yards across, with the largest ones in the center and smaller ones around the perimeter. They swarm with carved images of lions, foxes, gazelles, snakes, vultures, scorpions, and other creatures, as well as symbols.

When Göbekli Tepe ("Potbelly Hill") was first uncovered in the 1960s, people mistook it for an insignificant medieval graveyard. Thirty years later, Göbekli Tepe was called "the most important archaeological dig anywhere in the world." Built around 9500BCE, Göbekli Tepe is the oldest human-made place of worship. It was erected so long ago that the people who built it used *stone* tools—hammers and chisels made of flint, not metal—to hack the pillars out of the living rock. These enormous rocks were then transported to the site from a quarter mile away and erected, all before the invention of the wheel or the use of domestic animals. Göbekli Tepe was built by a small group of hunter-gatherers seven thousand years before Stonehenge, and eight thousand years before the pyramids.

When the first *Homo sapiens* looked out over the African savanna almost two hundred thousand years before Göbekli Tepe, there were still two other human species living on the earth: the Neanderthals in Europe and *Homo erectus* in Asia. Archaeologists cannot distinguish between the tools made by these three types of humans. We fashioned the same stone choppers, axes, and flakes as our contemporaries—the same tools that had been in use for the previous million and a half years. As far as we can determine by looking at fossils, these early *Homo sapiens* were physically just like us. Their bone structure is almost identical to that of your current neighbors, albeit a bit heavier and thicker as a result of a more difficult life.

All three species would have been incapable of the imagery, workmanship, and cooperation that went into Göbekli Tepe. But then something happened fifty thousand years ago—the most important event in human history. Even though the bones of humans didn't change, there was a sudden and remarkable explosion in the number and kind of tools these humans created. Our *brains* had changed; our ability to think exploded.

We know this happened because of the proliferation of artifacts: projectile points, knife blades, engraving tools, and drilling points. There were several advances in the manufacture of flint tools that led to smaller, finer, and much more useful implements, such as a burin, a kind of chisel. With an old stone hand axe, you are limited to very rough smashing and chopping. But with these tiny, delicate burins, it is possible to cut and carve bone and antler with precision.

At our first appearance, *Homo sapiens* had an energy budget similar to that of *erectus* and the Neanderthals. All were living in the old hunter-gatherer way, using the same stone hand axes, and consuming a similar diet. The revolution in thinking that created this new storehouse of tools slanted the energy equation strongly in favor of modern human beings. Suddenly (in evolutionary terms) humans had the means to greatly increase their energy intake while expending much less energy.

The Neanderthals, for example, living in freezing cold glaciated Europe, needed as much as 4,000 calories a day to survive because they had to burn the energy of a track and field star training for the Olympics to get their food. We live on about half that amount of calories and a wide variety of energy sources. Plant food was scarce in their icy world, so the Neanderthals hunted the hairy rhinoceros and woolly mammoth that roamed Europe at the time. Their spears were not meant for throwing because the stone tips shattered too easily. Instead they were thrust like a bayonet at close range. Imagine for a moment getting close enough to an angry rhinoceros to jab it with a stone-tipped stick. Neanderthal bones show many healed fractures: they were broken in the process of securing a meal.

When modern humans pressed into Europe, they had the advantage of a much better set of tools. They had a marvel of technology known as an *atlatl.* This throwing stick is a short piece of wood with a little curved notch at the far end. You place the butt of a spear in the notch and fling it, much like a lacrosse player uses a stick to flick a ball. The spears were tipped with another new invention: points made from antler or bone that wouldn't shatter on impact with the ground, as a stone point did. Human hunters using throwing sticks could stand at a safe distance from big game. These advancements meant gathering energy was much safer and more efficient, conserving our own energy for other things: thinking, imagining, conceiving, planning, manifesting. Creating an unimaginable complex of megaliths at Göbekli Tepe, planning 9/11.

And in Asia, we developed the tools for fishing. In 2001 a skeleton was discovered at Tianyuan Cave near Beijing dating from around the same time, forty thousand years ago. Collagen analysis of the bones reveals that this individual ate freshwater fish regularly. Some scientists think that the docosahexaenoic acid (DHA) molecule present in fish fats helped spur even more brain growth in humans. DHA plays an unusually large role in human brain chemistry because our brain cell membranes are made of fat. The gray matter, where most of our thinking takes place, is rich in Omega-3 fats, or DHA, because this chemical makes for healthier, more efficient neurons. Your body cannot manufacture DHA; you have to ingest it whole from a food source, and this molecule is not abundant in savanna animals. If DHA did indeed impact the development of our big brains, our evolutionary leap could have first occurred at seaside locations among the first fishermen, whether they used spears or hooks or caught fish with their bare hands.

The co-evolution feedback loop was in full swing. The ability to create tools meant better sources of energy, which meant more brain development, which meant better tools, which endlessly created more skill and more intelligence. Every object in the world had the potential to become something else, conceptualized first in the mind, manifested next with hands and the cooperation of other individuals. The rapid advances in tools and the expansion of intelligence helped to give birth to the conceptual revolution. And right in there with them was another new faculty: language.

***

A father and his eighteen-month-old daughter are outside on the seashore at sunset. It has been a long fun day at the beach, and mother is resting on a blanket. The little girl knows a few nouns and verbs but has never made a sentence. Most communication still takes place in the form of crying, yelling, laughing, pointing, stomping. Who is she? What does she think? The father can only guess.

The full moon is rising and he points to it, saying, "Moooooon." His daughter repeats the word, enchanted by his reaction. Filling his voice with wonder and awe, he explains how the light of the sun is reflecting off the moon, making it appear so bright. The little girl picks up on the tone and points at the moon, wide-eyed. Then she opens her mouth and says to him, "Mama look moon."

Call it magic or call it nature, the little girl has just taken her first step across the divide that separates her from Neanderthals and *Homo erectus.* In a single sentence she has spanned the million and a half years between the beginning of inarticulate human sound-making and articulate speech. She can communicate her mind, her intentions, her desires, what she wants for another person. Language takes off when we can combine isolated words into meaningful groups that communicate complex ideas.

Early *Homo sapiens* probably did not have this ability. Like a small child, they could say single words and make expressions, gestures, and grunts. Sentences make all the difference. Language becomes a tool for translating thinking into a form that can be shared. Once we could exchange our thoughts with others in a specific and unambiguous manner, our ability to work as a team blossomed. We could describe things we had seen, share knowledge with the group, coordinate team actions, provide warnings and incentives, and create friendships and alliances. We could externalize and share plans with our group and work together much more effectively. Language evolved very fast, blanketing the human population in only a few thousand years—a blink of an eye in evolutionary time. It has leveraged our capacity for social interaction and community effort ever since. It also meant our groups could remain stable at a much larger size.

Chimpanzee populations are limited in size to about fifty individuals because their grooming and bonding rituals, necessary for keeping the peace, are extremely time consuming: picking the lice from each other's fur and a lot of soft touching and petting. Grooming is how they establish "friendships" with each other. It takes up about 20 percent of their waking hours. The rest of the energy budget requires them to forage for food, care for the young, and mate.

This grooming requirement imposes a practical limit on chimpanzee groups because each individual has to maintain a relationship with all the other individuals. If there are too many strangers in the group, the necessary friendly bonding emotions aren't present, and tensions will cause the group to break in two. One chimp can create a grooming relationship with fifty others, but when there are sixty-five or eighty others for this individual to generate a friendly relationship with, there is just not enough time in the day. Imagine what would happen with a hundred or more individuals. There are too many strangers, too many chimps that have no connection to each other. The glue of positive bonding that holds the group together cracks, and the chimps will go their separate ways in smaller, more manageable groups.

Human beings' need to bond as a group was much more important than chimps'. Our hunting, gathering, child rearing, and other activities were all strongly dependent on group cooperation. Ruffled feathers, hurt feelings, and other emotions could easily spiral out of control. Humans groom too: think about how good it feels to have your hair fussed over, your nails cleaned, your shoulders massaged, or to be gently touched by another. It is profoundly soothing because it activates the parasympathetic nervous system—responsible for slowing heart rate, lowering blood pressure, deepening the breath—and has always functioned as a way for us to bond, calm, and restore peace to the group.

When speech arose in human beings it overcame the time-grooming hurdle in human populations. Connecting with each other, calming down, tending, and befriending are possible through language, and language functions much more rapidly than touch. Also, many people can share a conversation at once, multiplying the effects. And you can talk while attending to other tasks. A friendly word, a quick negotiation, a verbal gesture of support can defuse a potentially fractious situation in a matter of seconds. The time people spend in verbal social interactions worldwide is about 20 percent—the same as our chimpanzee cousins spend on grooming. But because language is so much more efficient than grooming for bonding, that same 20 percent can affect many more individuals, increasing the size at which human groups can be stable.

Scientists have discovered a single gene that made humans capable of modern language. Evidence shows that while it was present in a stable, rudimentary form in every mammal for millions of years, this gene began radically and repeatedly updating itself right around the time of the explosion in thinking that created all those new tools. Our new, large, DHA-marinated brains were imagining scenarios, making new predictions, and coming up with new projects like Göbekli Tepe. Every human community needed better communication to stay viable, and those without it were left behind: Neanderthals and *Homo erectus* disappeared at this time. The upgraded human species was quickly taking over the earth. Not only did communication create better cooperation; it helped us develop the ability to see into each other's minds.

***

Let's say you show some three-year-old children two dolls, one called Sally and one called Anne. The dolls live in a dollhouse, which contains a basket and a box. Sally has a marble she puts into the basket just before she goes outside the dollhouse for a walk. Anne takes the marble out of the basket and puts it into the box while Sally is gone. When Sally returns from her walk the children assume Sally will look for her marble in the box. They cannot understand that Sally doesn't know what they observed, that Anne moved the marble while Sally was out. To the children, their minds and Sally's mind are the same thing. They cannot conceptualize the mental state of a separate individual because their brains have not yet developed the capacity.

Around age four it becomes obvious to most children that Sally will look for the marble where she left it: in the basket. Even though the child knows that the marble is in the box, they can now build a model of Sally's mental state and understand that Sally doesn't know what they know, that Sally's view may be different from their own. Some people are better at this than others, but most of us do this in trivial ways all the time. Step on somebody's toes? It's easy to predict that they will be surprised, hurt, and irritated. Give them a hug and we know they will probably smile and forgive us. This ability, called theory of mind, took about four million years to develop.

Theory of mind means that we imagine, in our own minds, that other people have thoughts and feelings that are different from our own. When we imagine giving a child a piece of chocolate, we can also imagine that the child will be happy about the gift. That is a feat of theory of mind. When Tiger Woods hid his infidelities from his wife Elin, it was because he probably imagined that she would be upset if she found out about them. When terrorists plotted the 9/11 attacks, they imagined how that would affect

Americans, as well as Muslims. And Terry Jones accurately predicted that his Qur'an-burning stunt would attract worldwide attention to his little church. That was the human capacity for theory of mind in action.

Imagine yourself as a *Homo erectus,* trying to understand another *Homo erectus.* Her face is a mask of fear; she is gesturing forward and making sounds indicating she's scared. Looking around, you can't see anything obviously dangerous, but you get the idea that she is afraid and upset, and that whatever it is, it is over in the direction she is pointing. Your course of action might be to coax her to run away, to follow you away from the danger. That is about the best you can do.

Now imagine the same scenario, but occurring between two *Homo sapiens.* You can ask her what is the matter. She can tell you, "My child is injured down by the creek. Can you please come help?" On top of your empathic connection and the use of language, you can also use theory of mind to generate a detailed impression of exactly how she must be feeling and why. Instead of running away, you realize that the right thing to do is to follow her down to the creek to help her child.

Such precise, intimate mapping of other people's mental states was the secret sauce that allowed humanity to come together in large groups. We could imagine other people's inner worlds much more effectively and use this to behave in ways that would be more copacetic for everyone. It was like empathy on steroids, and together with language, theory of mind paved the way for large group endeavors.

Once human beings get the hang of imagining minds, there is no stopping us. Children attribute minds to almost anything, whether it is a person or not. To a child, a cloud can have feelings and intentions; a mountain can be sad or want a cookie; a toy car can be a best friend who understands everything about us. Grown-ups, too, see intentionality and feelings in inanimate things, although not usually with the literal sense of a child. Adults tend to attribute the quality of having a mind to intangible concepts such as spirits, gods, or devils.

In fact, evolutionary psychologists hypothesize that religion itself may have come about because of theory of mind. Our ancient human ancestors, endowed with theory of mind, began to see intentionality in nature, in weather, in the sky, and in concepts like the spirits of deceased ancestors. Obviously, these things—whatever their ontological status—do not possess literal, physical brains, but the theory of mind capacity in human beings did not evolve to connect the having of a brain to the having of a mind. It sees minds wherever it looks, and this is probably how the first spiritual concepts were born. The volcano erupts because the gods are angry. We are eating well this winter because the deer spirit is pleased with us. This capacity allowed us to come up with the concept of the mind of God and begin to speculate about what

God wants and how God feels about something. Thus religion was born.

***

It's easy for small groups of related individuals to engage in empathy and altruism because they share a large number of the same genes. But in big groups of hundreds of people, it's rare that everyone would be closely related to you. As humans gather together, the challenge is always how to trust people who are not our relatives. In all previous history, this would have been unthinkable; a stranger was an enemy. Helping strangers thrive did not benefit our own genetic lineage, and any humans predisposed to such behavior would eventually be weeded out by evolutionary pressures.

For a large group to function properly and thrive, people have to trust their neighbors. In fact it's a biological imperative since a large and diverse gene pool produces the strongest offspring. Language and theory of mind help us bond with strangers. But for a large group to function we must have a reasonable expectation that all the members of the society are starting with the same set of beliefs, the same values. We have to trust that others will pull their load, contributing equally to the group. We have to trust we won't be harmed by other members the minute we turn our backs. We have to trust that if we help someone, the exchange of energy will be a two-way street. Without this energy exchange the benefits of membership in the group are lost. It turned out that religion was the perfect solution to the problem of creating social trust. The bond of common beliefs, cultural ideas, or memes, became a social substitute for genes. Shared memes gather people into a conceptual family, and religious beliefs are memes in their most potent form.

The earliest forms of religion were shamanistic, in which animals and spirits were considered to have humanlike minds. Shamanic rituals created strong bonds of trust between the group members. Members of the faith swore oaths to the deity and to the group, and such oaths infused a deep sense of obligation to do right and shame about doing wrong. They agreed to codes of behavior that all would adhere to. The bonds of trust forged were nearly as strong as genetic bonds, making it possible for members of the society to work together as an extended family.

The theory-of-mind feeling that there is an invisible spirit who has a humanlike mind and is capable of watching you all the time made following these codes much more likely. Even dogs and cats understand that they can get away with things when their human companion is not watching them. They wait until we go to work and then take actions they know are forbidden. The concept that there is an invisible spirit that can watch you every second of every day meant that people felt truly afraid to transgress group agreements. The rules of fairness, trust, and helping others made the group cohere.

***

Emotions helped us become successful creatures, yet we see they also have a hidden cost. The same is true for thinking. Conceptual ability brought with it a host of new mental problems. For the very first time, we human beings could imagine dying as well as the death of our friends and family. No other animal, not even earlier human species, evolved to cope with this knowledge.

The inevitability of death is terrifying, and conceptual thinking provided some answers. We came up with ideas that turned into stories about how the world worked, that explained what happens to dead people, and that showed us what we could do to ensure we lived forever in another world.

As our anxieties multiplied, our conceptual world grew larger and more complex to deal with them. Why do people get sick? What is birth and life? Why do people die? Where do they go? How can we make sure the hunting season will be good? As people thought about these questions, they shared their stories and answers with each other, creating a common culture and religion. For the first time our behavior was determined not just by evolutionary pressure. We used ideas and language to agree with each other about how we should act. And we were sure these ideas were deeply connected to our survival.

Because we now had shared ideas, we could conceive of large projects and manifest them. The earliest evidence of this is at Göblekli Tepe, which might be the world's first temple. There, two tall pillars in the center of each circle of the megalithic rings are carved with semihuman forms, the first depiction of spiritual beings. Their heads have no eyes, no mouths, or faces. But they have arms and they have hands. They are makers.

***

Can you imagine a world without imagination? It is literally impossible. The conceptual revolution that occurred fifty thousand years ago fundamentally changed our experience of being human. Now we could imagine each other's minds and feel what another person might feel. We could ask them about it and receive a reply that helped us understand more clearly and deeply. We could dream about the future and ponder the past. Our ability to think was just taking off, and humanity was beginning its boldest experiments. Religion, culture, art, tools, construction, communication: it was all novel and crackling with frisson.

The marriage of thoughts and feelings into beliefs not only brought us together; in a larger sense it forced us apart. Each community had its own language, its own culture, its own set of beliefs, and all were held equally dear. Because of negativity bias, we felt afraid of foreign beliefs. We could take new kinds of actions: attacking a *symbol,* like burning a Qur'an, causing real distress. For the first time we could experience religious or cultural hatred and direct that emotion at an entire people. Hooking our group identity to beliefs has powerful advantages; it keeps us together and gives us a sense of purpose. And it became part of the evolutionary feedback loop that has brought us to where we are now: trapped in the prison of thoughts and feelings, in a dangerously unbalanced relationship to mental concepts.

chapter 8

prima donna

Mohammed bin Laden, born penniless in Yemen, became the illiterate founder of an enormous multinational construction conglomerate and holding company. In his lifetime, the Saudi Binladen Company rebuilt the three most holy sites in Islam: Al-Aqsa Mosque in Jerusalem, the Prophet's Mosque in Medina, and the Grand Mosque in Mecca. Mohammed bin Laden fathered around fifty-five children, including Osama, child number seventeen. Osama's mother divorced Mohammed soon after Osama's birth, and the boy was raised from infancy by his mother and her second husband in the seaside port of Jeddah, gateway to Mecca.

Osama was a tall, willowy, shy youth, who spoke softly and covered his mouth when he smiled. After a religious conversion at fourteen, he became a very serious young man, prone to the sort of extremism typical in a teen. His mother was concerned, but she soon saw that Osama was set on a pious life. All his brothers went off to school in wild, Westernized Lebanon, where they were introduced to every kind of mischief from drinking to sex, from American culture to rock music. But Osama stayed in parochial Jeddah, attending the prestigious King Abdulaziz University. Although his grades were middling, his heart was stirred by the rhetoric of some of his professors, who were Islamic fundamentalists on the lam from secularized dictatorships in Egypt and Syria.

One of these professors was Mohammed Qutb, the brother of Sayyid Qutb, whose ideas are at the heart of modern Islamic terrorism. Osama went to Mohammed's weekly talks and was fascinated by the radical form of political Islam that promised to return the whole world to righteousness. Mohammed Qutb, like his deceased brother, had endured the Egyptian torture cells for the sake of his religion, and to Osama and his classmates he had the romantic aura of a spiritual hero. He taught that Islam was not only a religion but also a political philosophy, and advocated a return to a worldwide fundamentalist Islamic theocracy. Anyone who got in the way of this vision was not a real Muslim and could be killed with impunity. Osama liked what he heard.

At seventeen Osama married a fourteen-year-old girl from his mother's village. Their wedding was an austere affair: no dancing, singing, or even laughing was allowed. Saying that Islam needed many warriors for the *jihad,* bin Laden soon married three more women. His home was forbiddingly bare. No telephones, television, or air conditioning were allowed, and there was almost no furniture. Toys were barred from the household. Though he came from one of the richest families in the world, bin Laden idolized the poverty of the saints and the illiterate savvy of his father.

To toughen up his growing brood, he took the children to a rustic farm where they would sleep outdoors in the sand without blankets. He kept them out of school and hired private tutors to ensure the purity of their education. When his second son was born with hydroencephalitis ("water on the brain"), bin Laden took him to a specialist in London but refused to allow the surgeons to operate. Instead he whisked the boy back to Jeddah and treated him with traditional remedies such as honey and prayer. The boy lived but has suffered the consequences of serious brain damage that could easily have been avoided.

Bin Laden was enraged when the Soviet Union invaded Afghanistan in 1979. He felt the conflict was symptomatic of the godless West's desire to destroy Islam. He dropped all other concerns and turned his house in Jeddah into a way station for fighters bound for Afghanistan. Using his connections among the wealthy elite of Arabic society, he raised tens of millions of dollars to support the mujahedin. As his support grew he moved to Peshawar, Pakistan, just over the border from Afghanistan. There he set up a guesthouse that soon filled with idealistic teenaged Arabs, on fire with the dream of defending the faith and martyring themselves in combat with infidels.

Nearly five years into the conflict, bin Laden had gained a reputation as a fighter back home but had never actually set foot in Afghanistan. Eventually he was coaxed over the border, where he witnessed the bravery of the Afghani mujahedin under a bomb attack by Soviet jets. Convinced cover was necessary, he set about doing what he had learned to do by watching his father: organize construction. Gathering heavy equipment from his family's business, he built the Tora Bora cave complex in the mountains just inside Afghanistan. He moved his untrained army of restless teenagers from Peshawar into this bunker but refused to allow them to fight. By the time the Russians left Afghanistan, bin Laden's men had only engaged in one significant encounter in which they drove off an attack on the complex by a small group of Soviet paratroopers.

The encounter meant nothing to the Russians, but in bin Laden's mind his ragtag band of Muslim holy warriors had driven a spear into the heart of a superpower; it was the blow that brought down the entire edifice of the Soviet Union in 1989. Faith, piety, submission to Allah, and a fierce desire to fight for righteousness could overcome any worldly force. He kept a trophy from his encounter with the Russian paratroopers, a small AK-74 assault rifle, which was with him when he died.

Bin Laden didn't set out to become an international terrorist and mass murderer. All the descriptions of him were of a gentle, kind, and thoughtful young man. Over time his identification with Islam, and his feelings for what he saw as the downtrodden and oppressed Muslims of the world, began to move him toward an ever more radical philosophy.

His image of himself evolved slowly, first as a pious father trying to raise his children right. As the Soviet war developed and he raised millions of dollars for the Afghan rebels, he saw himself as the defender of the faith against godless communists. And when he got personally involved in the fighting and saw the Soviets retreat, he came to believe that he was a savior, the sheikh who would bring about a vision of Islamic world domination. It was then, swollen with this grandiose vision of his own significance, that the ratty, brokedown philosophy of murdering innocents made sense to him.

***

Who you think you are is subject to change; to being reexamined, altered, updated, edited, thrown out, rebuilt, and renewed. When you get married, you become a wife or a husband. When you have children, you become a parent. You may have been a widower, divorced, a CEO, a writer, a person who just got a job, a person who just lost a job, a new home owner, someone who lost a home in a flood. You've been a child, a teenager, and now an adult. The impositions of circumstance, the twists and turns of fate, all of these are occasions that bring about a big moment of self-reflection: who is this being that you are constantly redefining?

We think we know. We wake up every morning sure of the sense of who and where we are. We go through our day feeling solid about who we are. But what is "me"? The body may be the element we think of first. We may even point to the heart when we say "me," but is the body what we refer to when we ask the question?

The conceptual revolution gave human beings the capacity to imagine things in great detail, to build virtual worlds in the mind. There, we play out scenarios. We speculate about the future: if I don't ever go to New York or Jerusalem, maybe I'll never experience a terrorist attack; if my daughter gets into the state college, I can buy her a car. Or alternate histories of the past: if I had taken that job in Pasadena, I would have reorganized the company and my career would have been in a much better place by now; if I had been tougher with my son, I would have placed more emphasis on his homework and taken away his video games and he'd be better in school. We imagine ourselves speaking at tomorrow's meeting, driving home, eating, and watching a movie at the end of the day.

If you are planning to cook a meal, you need to imagine the grocery store, the groceries, the cooking utensils, and the stove involved. And you need to see how all those items are going to come together in a sequence that will produce a tasty meal. But without the main character who is gathering the utensils, reading the recipe, going to the grocery store, the plan couldn't come together. You need to see, in your mind, the person who is going to do all this stuff. This character in the imagination, in the mental representation of the real world, is a stand-in for ourselves. It is not really us, not our actual physical body and brain, but instead is an imaginary representation of us. That creation is what we label "me": the ego.

The mental self-concept, or ego, is a function that helps unify and coordinate our behavior and is at the core of our ability to plan. It is composed of certain types of mental images, such as pictures of our body, face, the house we grew up in, the schools we attended. It is also composed of ideas, such as our name, our ethnicity, our city or country, what type of person we think we are (tall or short; fat, muscular, or thin; smart or not), the many names people have called us, autobiographical memories, and ways we have dressed and looked. Gender, age, job title, social status, relationships with other people, political and religious affiliation, belief systems, all play a role. The ego identity of an adult human being is quite complex and includes a tremendous number of concepts.

All our capabilities have evolved out of biological need and come from skills and physical parts that already exist. Emotion evolved out of the pleasure/pain stimulus of the nervous system, which functions to protect the organism from harm and maximize the ability to survive. Concepts emerged out of the organism's need to make energy gathering more efficient. The mental self-representation evolved to facilitate complex behavior. Without a unified sense of self, our left leg might try to take us to the movies while our right leg went to work. One part of the brain might decide to plan making dinner while another part decided to go to sleep. There would be no coordinated action. Organization takes place on many levels, but at the top of the pyramid, running the show, is the ego.

***

We feel as if we have been the same person our entire lives. This continuity of identity is a natural impression, yet it is demonstrably untrue. As children, for example, the stories in our heads are much simpler, and the mental images of our bodies entirely different, as suits our smaller bodies. As new experiences build up a richer life history, the memories of this history are added to the system, changing it. Our physical body undergoes various transformations over the course of a life. When we are children our teeth grow in, drop out, and get replaced by a second set. In our teen years our sexual features develop, and new patches of hair spring up in our pubic and armpit regions. Young women begin the cycle of the menses and young men grow beards. When we are elders our bodies change yet again, becoming weaker, our hair turning gray. Even our face, that icon of identity and selfhood, is always changing, becoming wrinkled, sunburned, scarred, and altered by time.

These physical transformations are echoed by changes in our social role and standing. As toddlers our social world is mainly composed of our family members, but when we go to school we find a wider identity as a student among other students. Here we come to think of ourselves as smart or slow, social or introverted, popular or not, a good student or a behavior problem. Driven by potent hormones, the onset of puberty raises our social experience to a new level of intensity, and we find roles based on our emerging sexual identity, such as flirt, wallflower, bad boy. And when we graduate into the adult world, we again take on new forms of social identity, such as banker or janitor, homemaker, or jailbird.

Through all these transformations of life, our mental self-concept is constantly updating itself, changing, growing, and adapting to our condition and circumstances. Even though the alterations never stop, and we as an organism are in continuous flux, the mental self-concept *feels* as if it never changes. We experience ourselves as having always been the same person inside. This subjective impression of continuity points out the ego's purpose as a unifying function. If it didn't feel the same, it couldn't serve its function to create order out of the chaos of concepts, emotions, sensations, and experiences.

Unless there is a major psychological pathology, an adult human feels him-or herself to be *one person,* not many, and this conviction is central to his or her personality. It is not surprising, then, that for thousands of years the majority of theological and philosophical speculation has regarded the self as a singular entity, the very meaning of individuality and identity. Both Aristotle and traditional Chinese medicine placed the self in the heart. Descartes speculated that the self must be located in the pineal gland because this gland is a singular structure centrally located among the many duplicate structures in the brain.

Today virtually all major researchers in neurology agree it is very unlikely that the ego is located in a single brain system. The ego arises from the combined output of many different modules located in different places throughout the brain. Various theories model the self-concept as a network of associations or a reflection of sensations based in body awareness and emotional tone: The pleasure/pain of sensations, the "I feel good/I feel bad" push-pull of emotions. Joseph LeDoux's theory of the "synaptic self" goes even further in this direction, stating that the sense of self arises ubiquitously in the synapses of the brain, virtually everywhere. There is no ego module, no self spot in the brain.

These different theories all agree that the mental self-concept is a composite of many different elements. It is made up of an interdependent interplay of body sensations, emotions, thoughts, memories, and plans. And if you examine each element, there is no end to the division of them: nerves are made of cells, which are made of organelles, which are made of molecules, which are made of atoms, which are made of subatomic particles ... and there is no final Russian-doll particle to which we can point and say "this is it, the bottom line." Similarly, each thought has a stimulus, a story, and a result. Each memory melds with other thoughts and if you try to trace it back to a single isolated point in time, you'll find that it only exists in the context of other memories and thoughts; there is no original thought, sensation, or experience you can point to and identify as the beginning of everything. It's as though the sense of self, the ego, will fly apart upon examination, and nowhere do we find a self-entity, an eternal and unchanging essence.

Even though the ego is ultimately indefinable, we still experience ourselves as a unified whole, reducing the totality of who we are into this composite entity that becomes encapsulated into an idea in our head. This "idea of me" then becomes our identity. The word "identification" derives from the Latin word *idem,* which means "same" and *facere,* which means "to make." We make ourselves into an image of who we think we are, a mental picture of ourselves, backed up by a host of self-referential thoughts and feelings.

We are not born with this mental self-concept. It develops early in life just as it evolved early in the conceptual revolution. An infant has no ego. If you place him in front of a mirror, he will act as if he is encountering another baby. He will paw at the glass and crawl behind the mirror trying to find the mysterious other. But around eighteen months he suddenly has quite a different reaction. He will look at the image in the mirror and then point to himself, or even say "me." The light has gone on. He has developed the idea of "me," the genesis of an ego.

Several animals besides humans can recognize themselves as a reflection in a mirror, including adult chimpanzees and orangutans, suggesting that they, too, have the rudiments of a mental self-concept.

Previous to the conceptual revolution, human brains could not generate a complex ego. The archaeological record speaks volumes about the immense change the ability for self-reflection brought about. Suddenly there was true art, with elaborate, intricate depictions of people and animals, some of them altered in imaginative ways. For the first time there was an explosion of ritual or cosmetic adornment of the body: shells, paints, feathers. In place of practical disposal of corpses, the dead were ritually buried with objects and adornments, and in symbolic postures. With imagination came the ability to imagine a much more complex and fortified ego. And as the capacity for language increased, people began telling stories about this new experience of self.

***

I will proclaim to the world the deeds of Gilgamesh. This was the man to whom all things were known; this was the king who knew the countries of the world. He was wise, he saw mysteries and knew secret things, he brought us a tale of the days before the flood...

When the gods created Gilgamesh they gave him a perfect body. Shamash the glorious sun endowed him with beauty, Adad the god of the storm endowed him with courage, the great gods made his beauty perfect, surpassing all others, terrifying like a great wild bull. Two thirds they made him god and one third man.

Gilgamesh was an actual human king of Uruk (a city-state in what is now Iraq), who ruled around 2700BCE. He built the walls of the city and refurbished the temple of the goddess Ninlil. As king of one of the first city-states, he became the center of a whole mythology, and his fantastical adventures of slaying beasts with Enkidu and racing the sun constitute our first literature. In seven thousand years we went from the faceless anonymity of the makers of Göbekli Tepe to the ego of a single person as the hero of its own mythology.

The formation of cities was the cause. As people settled down, almost two million years of hunter-gatherer egalitarianism came to a crashing halt. Staying in one location meant that a person could amass property and wealth and that some people could accrue more possessions than others. Religion created a strong social contract and brought people together to make large projects, so it's likely that the first leaders were priests. Then as now, religion could be made to bring rulers more wealth, legitimize their position, and compel others to do their bidding.

With the ability to organize others *en masse,* these leaders learned to defend their possessions and take the possessions of others. Raiding parties became common, as evidenced by the massive stone walls that turned these little towns into armed fortresses. One of the first of these was Jericho, a city that still exists after more than eleven thousand years of continuous habitation. Its population of over a thousand individuals was surrounded by a wall about twelve feet high and three feet thick, and included a tower more than twenty-five feet tall, weighing an estimated thousand tons. Such massive construction projects took organization, food, and rules, but most of all they required the unified, singular vision of a leader to bring everything together.

Religion and military power colluded to elevate the leaders of cities like these into god-kings. Considered to be divine beings, they commanded total obedience from their subjects and gave birth to the basics of statecraft. The spiritual authority of the ruler became the basis of civilizations. Even into the early twentieth century, Christian kings and Muslim sultans were thought of as God's agents on earth. The last Chinese emperor, Puyi, was still called the Son of Heaven in 1922; and the current Emperor of Japan, Akihito, is to this day referred to as the Tenno, or Heavenly Sovereign.

The first god-kings levied taxes to pay for the weapons, armor, and feeding of their armies. Entire classes of civil servants came into being for the first time—viziers, tax collectors, soldiers, and scribes—a trebling of social complexity. This was such a successful way of organizing people that cultures grew until they were made up of many such cities, like the sixty cities and 2 million citizens making up the empire of Egypt. The pyramids—which, after all, are nothing more than enormous mausoleums for the god-kings—are monuments to the colossal ego of the pharaohs: testaments to a time when tens of thousands of people were slaves to the will of a single individual whose personal story was woven into state mythology.

***

As the leader of a terror organization, bin Laden saw his story grow in a similar manner. Arabic language newspapers trumpeted apocryphal tales of his struggle against the Soviet army. Attributing miracles to himself, bin Laden said, "The Russians were trying to capture me, but I was so peaceful in my heart that I fell asleep. This experience has been written about in Islam's earliest books. I saw a ... mortar shell land in front of me, but it did not blow up. Four more bombs were dropped from a Russian plane on our headquarters, but they did not explode." Stories like this may be puzzling to non-

Muslims, but to the faithful they contain resonances with the stories of the Prophet Mohammed.

Before he left Afghanistan, bin Laden founded an organization meant to carry on the work of jihad into the future. Called simply The Base, *(al-Qaeda)* its goals were to train mujahedin to fight for Islam, "...to lift the word of God, to make His religion victorious."

On his return to the Saudi kingdom, Osama bin Laden was heralded as a hero of the Afghan war, but his good relationship with the government did not last long. In August of 1990, Saddam Hussein invaded Kuwait, embroiling Saudi Arabia, which shares a border with both Iraq and Kuwait, in a deadly conflict. There was real concern that Hussein's mighty army would roll over the border and capture the Saudi oil fields too. Bin Laden begged the Saudi government to allow him to raise an army of mujahedin and unemployed local youth with which to smite Saddam. When Saudi officials confronted bin Laden with the reality of Saddam's million-man army, its heavy tanks, and cadres of artillery, jet fighters, and Scud missiles, Osama replied, "We will fight them with faith."

The Saudis rebuffed bin Laden and opted for protection from the Americans, who put half a million American troops on the ground in Saudi Arabia within weeks. Bin Laden was livid when he saw the ranks of infidel "crusaders" darkening the holy land. Even more upsetting was the fact that some of the foreign military personnel were women, which bin Laden, and most Saudis, considered not only sacrilegious but a grave insult to the manhood of his people. "Women! Defending Saudi men!" bin Laden railed. One government minister who knew bin Laden was shocked by the change in him. He had gone from a quiet, humble, pious man to a person who seriously believed that he could raise and command a holy army to defeat Saddam. Not quite ready to declare bloody war on America—the United States had been his ally in Afghanistan—bin Laden nevertheless called for a boycott of American goods, a kind of economic jihad.

His strident criticism of the Saudi government and funding of political killings across the border in Yemen eventually got him kicked out of the country. His 1992 exile took him to Sudan, just a few hundred miles across the Red Sea from Jeddah. In the capital, Khartoum, he quickly erected a new base for his mujahedin.

Denied the opportunity to train an army of volunteers to save his country, he instead began training his own children. Taking his four wives and fourteen children out into the desert, he made the older boys dig foxholes for all of them. Even the toddlers were made to lie in the foxholes in the freezing night. "You must be gallant. Do not think about foxes or snakes," he lectured them. "Challenging trials are coming to us," he said, meaning the coming war with the West. When the children complained of the cold, he told them to cover up with dirt or grass, saying, "You will be warm with what nature provides."

This was not their only trial. When his driver ran over a monkey his kids loved, he told them that it "was not a monkey at all, but was a Jewish person turned into a monkey by the hand of God." And to test the poison his mujahedin were experimenting with, he gassed a litter of puppies his boys had adopted.

It was in Sudan that bin Laden reformulated al-Qaeda's mission. In his eyes the Soviet Union was no longer the biggest threat to Islam; instead it was America that stood in the way of a global Islamic renaissance. Rather than defending Islam, al-Qaeda would go on the offense, a global terrorist organization dedicated to attacking American and Western influence wherever he, bin Laden, saw it.

His first actions were to fund small bombings in Yemen and Somalia, and he made the dubious claim that his men were behind the "Black Hawk Down" incident. But then he got involved in a botched plot to kill Egyptian President Hosni Mubarak in 1995. Bin Laden was implicated, lost his Saudi citizenship, and, much more importantly, was cut off from his family fortune. Overnight he became a destitute and stateless renegade.

Embittered but determined, bin Laden returned to the only country that would have him, Afghanistan, under its new Taliban rulers. In 1998, he and his friend Ayman al-Zawahiri, a cofounder of al-Qaeda, signed a *fatwa* called the "World Islamic Front for Jihad Against Jews and Crusaders" that proclaimed killing Americans and their allies to be the duty of every Muslim. "[Americans] are very easy targets," he said publicly. "You will see the results of this [fatwa] in a very short time." The shy young student, concerned with modesty and living a holy life, had just made a public pronouncement of his commitment to a global campaign of murder.

***

The ego is one of evolution's great masterworks. It concentrates the mind down to a small point of self-concern, bringing our survival mechanism to a new pitch of effectiveness. Essential to our survival, the ego represents something of a paradox as well. A human being without a mental self-representation would be nonfunctional in the modern world, unable to cope with the complexities and issues of life, and at the same time the self-concept can be the source of all our troubles. This imaginary entity is the cause of endless disappointment, insecurity, indignation, striving, shame, and conflict. Just like the pleasure/pain principle of all life and the emotional ups and downs of all mammals, the mental self-representation constitutes both a brilliant means of motivating and directing behavior and an imaginary problem that we are constantly trying to solve.

Many of the concerns and worries that make up the ego are, when seen from the outside, completely unnecessary. When any particular life issue is not just another difficulty to be overcome, but is instead *my problem,* it seems much more urgent and compelling and we allocate more brainpower and physical resources to solve it. The ability to model future scenarios in the mind becomes, with the help of the strong emotions associated with egoic concern, the self 's nonstop planning for its own happiness and success. The ego makes us take things personally. Just a concept in our heads, an evolutionary tool for maximizing the organism's chances for survival, it becomes the focus of all our concerns, the star of all our dramas, and it makes us strive harder and reach higher.

Does our egoic need for self-aggrandizement really necessitate all the suffering it causes? Do we really need to feel so anxious about whether everybody likes us or not? It is not for nothing that Eastern religions have declared war on the ego, singling it out as humanity's greatest source of suffering.

The ego seems to have tremendous value because it seems to be "me," the organism our nervous system is always seeking to aid and abet. If something contradicts its imaginary unity—like being passed over for a promotion, or being condescended to at home—we respond as we would to a physical threat, even though nothing but a concept is under threat. These types of emotions help congeal the mental self-representation into a feeling of being real, but the added kick comes when value and ownership are projected into the external world.

***

If somebody asked you to choose any letter from the alphabet, which one would you choose? It turns out that most people will unwittingly pick letters from their own name. We also have a subconscious preference for numbers associated with our birthday, and we like people who have the same birthday, have names beginning with the same letter, have the same careers, live in the same city, or come from our hometown. We often don't realize that we have made these distinctions at all, let alone know why. Research shows that most people cannot find a pattern in the choices they make, but the researchers know: the subconscious mind is responsible, and it has a strong bias toward liking things with which we are identified.

Psychologists have developed an ingenious way to find these subconscious identifications: the implicit association test (IAT). Implicit association means things that are linked in our minds that we are not consciously aware of. An IAT uses implicit priming—subconscious suggestions so subtle that the subjects don't notice—to discover these mental associations. Implicit primers could be something as simple as which letters are used more often in a paragraph or numbers that appear more frequently in a list. The subconscious mind picks up on these patterns and begins to make meaning out of them beneath the level of conscious awareness. When a researcher manipulates this pattern-recognition capacity on purpose, it is possible to discover and actually measure our subconscious biases.

Taking an IAT usually involves matching groups of words as fast as possible. If you make a list of words like Lincoln, democracy, Constitution, and Washington, Americans will associate these much more quickly with positive words like good, best, excellent, and superb than they will be able to distinguish them from negative words like bad, worst, terrible, and awful. On the other hand, if you test the same Americans on a list of words like Mohammed, Qur'an, Islam, and mosque, they have the opposite difficulty. It is much harder to see them associated with positive words. Tests like this can make concrete measurements of our implicit biases, showing us the secret attachments that anchor the ego in identification.

We have a genetic predisposition to bind ourselves to our tribe. Religious, racial, and nationalistic attachments are particularly strong. Often these are drilled into us from birth, so these neural structures likely underpin much subsequent learning. Many people who think they are free from racism actually test high on an IAT for racism. It's not that they are lying to themselves, it's that the biases are below the level of conscious awareness. They just don't realize that they are there. Some of these biases are very surprising—such as the fact that many American blacks test as biased against black people, and gay men are prejudiced against gay men—until you consider the fact that these attitudes are subconscious and implanted by the society around us when we are still very young.

Whatever we consider ourselves to be, whatever group we identify with, we carry a strong, subconscious preference for it because it makes up part of our mental self-concept. It is an aspect of the self; we have a strong attachment to whatever group we identify with, and a negative opinion of anything that is, even as an idea, against it or opposite to it. Our subconscious mind is always trying to build us up and break our "enemies" down.

Subconscious identification with groups can occur under the most minimal conditions. Scientists found that when participants were implicitly primed with pronouns such as "we," "us," "our," people responded to pleasant words faster than when primed with pronouns "they," "them," "theirs." "Our bus" is much better than "their bus" only because it is ours—something called the *endowment effect.* The mere presence of ingroup/outgroup association is enough to create an implicit bias about an otherwise neutral term. Even the price of a coffee mug can change: without ownership, test subjects said it was worth four dollars. With ownership, the price went up to seven dollars. Identification is that easy to create, and it becomes the conceptual framework that distorts reality and alters our neuronal connections.

Imagine a sports fan, Bill, who is strongly identified with the Denver Broncos. He rarely misses a game and goes so far as to wear a team jersey and cap while watching. Bill thinks of the Broncos as his team, often saying things like "We won today" or "When we get to the playoffs..." When the Broncos win a game, he is everybody's best friend. When they lose, he mopes around for days. People joke that they don't have to watch the news to see how the Broncos are doing; they can just take a look at Bill. Of the dozen or so games on any given Sunday, the only one that matters to Bill is the Broncos'. The outcomes of the other games don't affect his mood.

Bill's identification with the team is composed of a mental concept: an idea that he and the team are related. There is his self-concept and the concept of the team, yoked together by a strong emotion. In effect, the Broncos are part of his ego. This relatedness imparts an added value to the Broncos in Bill's mind. The Broncos' ups and downs become Bill's ups and downs through a subconscious process of identification. And like most forms of identification, his connection to the Broncos has no real-world meaning. He is not the actual owner of the Broncos, nor does he even bet money on the games.

The identification is mental, but the effects upon the physical body can be measured. One study focused on the testosterone levels of men watching a college basketball game. The game was extremely close, with the winning baskets coming in the last few seconds. Despite the narrow victory, scientists found that the men's testosterone levels increased by 20 percent—a significant amount—if they were on the side of the team that won. The fans of the losing team experienced a similar testosterone decrease. They even found that fans' testosterone levels rise *before* a game in anticipation, just as the actual athletes' do. The bodies of fans react as if they were physically taking part in the sport. Just take a look at a group of sports fanatics like Bill cheering on their team, jumping up and down, waving their arms, cursing the television screen, and slapping each other on the back—exactly like the team in the locker room.

The essence of the problem is that when we identify with something, it subconsciously becomes a part of our self-concept. We mentally assign it a connection to our own life. And when something happens to this conceptual entity in our mind, we react as if it were actually happening to us. A modern adult is actively identified with such a large number of things that we are almost constantly under some such "threat" or another. Our political cause flounders and we react as if our body is under attack. Someone contradicts our intellectual position on a topic we feel strongly about and we may feel so threatened that we imagine physically assaulting them. Poets have had fistfights, professors have slapped students. We may even feel that our religion—a concept in our head—is somehow under siege, and fly a jet into a building.

***

As Einstein wrote, “A human being ... experiences himself, his thoughts and feelings as something separated from the rest—a kind of optical delusion of his consciousness. This delusion is a kind of prison for us, restricting us to our personal desires and to affection for a few persons nearest to us.” The ego, formed to create order out of chaos, embedded instinctual principles for survival deep in the mind and heart. These instincts of identification with “me” and “mine” pass below the radar of conscious awareness and trap us in a defensive posture, ready for a fight.

The history of civilization is a natural history of the ego: dominance, control, power struggles. And it is the history of the rise of

humanity, allowing us, through its organization, to reach levels of unimaginable effectiveness. Trade, economy, writing, the wheel, domestic animals, architecture, math: the inventiveness of the new human brain seemed to know no limits. It lifted us out of our animal state and allowed us to reflect on our own experiences. We built ships and covered the earth.

And yet behind it all is a hidden limit, hidden in the way water is from a fish or air from a bird. We live with a conceptual self the same way we breathe air: it’s so pervasive that we don’t notice it or question its veracity. If you really think about breathing, you soon see that the only thing standing between you and death is the ability to breathe in. If you breathe out and can't breathe in, you die.

And if you really take a look at the ego, you might begin to question the motivation for everything you do: why are you running around trying to please a self-concept that appears solid but that you can't find, that doesn't seem to exist? It's hard to function day to day when these big questions are front and center.

A few human outliers had the presence of mind, the conscious awareness to see the hidden limit of living within the framework of a conceptual self and they began talking about it. It would take thousands of years before many of us, from the average person to Albert Einstein, would be able to make sense of their explorations. And in the meantime history, progress, and the conceptual revolution moved inexorably on, perhaps beyond the point where the ego has outlived its evolutionary effectiveness.

chapter 9

wake-up call

On July 4, 2006, the Space Shuttle *Discovery* blasted off and docked with the International Space Station. This 115th shuttle mission brought equipment to the space station, as well as supplies for the station crew. Mission specialists Lisa Nowak and Stephanie Wilson used the shuttle's mechanical arm to maneuver the massive *Leonardo* module, containing several tons of material, from the space station into the shuttle bay. Nowak also used a camera mounted to the robotic arm to inspect the shuttle for damage to its delicate tiles that may have occurred during liftoff. The wake-up call for day number four, "Good Day Sunshine" by the Beatles, was chosen by Nowak's family because of her sunny personality.

As a child Nowak had eagerly watched the Apollo moon landings and had dedicated her life to becoming an astronaut. After receiving an advanced degree in astronautical engineering from the U.S. Naval Postgraduate School, she specialized in electronic warfare in the navy. Then in 1996, her childhood dream had come true when she was selected for astronaut training and became a mission specialist in robotics. She trained for ten long years for her shuttle mission in 2006. It went off without a hitch.

Just a few months after returning to earth, in February 2007, Nowak drove east on Interstate 10 from Houston swaddled in a pair of astronaut diapers. She had packed latex gloves, a black wig, a BB pistol and ammunition, pepper spray, a hooded tan trench coat, a two-pound drilling hammer, black gloves, rubber tubing, plastic garbage bags, and an eight-inch Gerber folding knife. Her mission this time: get even with Colleen Shipman, the woman who had stolen Nowak's lover, astronaut William Oefelein.

A woman with a brilliant mind, Nowak was married and had three children, was internationally famous, and could look forward to continued success in the future. Her job was operating robots in outer space; she represented the absolute pinnacle of a modern human in a technological society. Yet the ancient hominid wiring in her brain remained much as it was in our prehuman ancestors: she felt jealous of a sexual rival and decided to remove her from the field of competition.

This type of behavior may seem extreme today, but as humans settled into cities ten thousand years ago, it would have been quite common. Hunter-gatherers in the field treated an unknown person as a hostile invader, an enemy to be put down or driven off. Killing and eating each other were frequent occurrences in our early days.

Human society had been getting more interactive, altruistic, and egalitarian since *Australopithecus,* but major behavioral shifts would have to occur in people who had been living by the spear for over a million years. Imagine life in a society that had not developed any manners, politeness, mediation, or other methods for smoothing things out. It would be like a crowd of unsupervised, xenophobic, three-year-olds bashing it out in grown-up bodies, having full permission to chuck their iPhones at each other during a meeting where colleagues disagree over how to take out the competition, and bashing them to death with the coffee machine after. We had to learn to live together without murdering each other over every dispute or sexual rivalry.

It is exactly this kind of mischief that societies like Göbekli Tepe, Uruk, and Jericho had to cope with, because the "wild" humans coming together in these communities had a propensity for violence. More than one out of four people died by murder in those days. Rules and codes were enacted to help stop the carnage, but the real change came from within, through the process of *human domestication.*

The first sign of a species becoming domesticated—more docile and less aggressive—is in the gracilization, or thinning, of the skull. The skulls of Cro-Magnons are much heavier and thicker than the skulls of modern people. In humans, gracilization begins to be noticeable around forty thousand years ago, right around the time of the explosion in toolmaking.

Learning to cooperate moved us away from our food-clothing-shelter survival mode and toward a much more creative, conceptual way of being. The change was slow at first, but there was a lot of low hanging fruit—basic inventions—for the human species to discover. We domesticated plants and animals and learned the ways of farming. Systems of irrigation were invented, which boosted production, and food storage methods came into being to keep the surplus from going to waste. Staying in one place meant that humans could amass food, tools, metals, animals, and other valuables, which allowed economies of trade to come into being. To protect these goods and keep the peace, laws of property and ownership, and the armed enforcement that go with them, arose. As government and political structures grew, we invented advisers, soldiers, tax collectors, and scribes.

Scribes started keeping records of each farmer's agricultural produce, probably for tax purposes. For thousands of years, writing was restricted to record keeping, until around 2600BCE, when scribes in the Sumerian city of Ur wrote the names of kings and queens on funereal bowls, and eventually prayers for them with sentences that included nouns, verbs, and all the syntax of spoken language. By 2000BCE this technology would lead to an explosion of literature, law, history, religious texts, and scholarly works. This was the time of contemplation, the birth of philosophy and the Axial religions, the era of Socrates, Buddha, Confucius, Mahavira, Elijah, Isaiah, and Zoroaster.

Technologies grew one upon the other: from writing came the symbolic language of math. From mathematics came everything from ways to track the movement of the stars to massive construction projects: river damming, irrigation, temples, and pyramids. The city-states traded, warred, and intermixed with one another, and gradually larger political units formed.

By about 3200BCE the thriving culture of the Nile river valley was united into a single kingdom with a population of 2 million under King Menes. From their new capital at Memphis, Menes and the kings who followed him controlled a massive labor force, the agricultural bounty of the Nile delta, and lucrative trade routes from Africa to Asia Minor. Egyptians made radical advances in art, architecture, technology, and administration. It would be another fifteen hundred years before cultures in China, India, and Mesoamerica would catch up to the advances in the cradle of civilization, the Middle East. By the reign of Augustus (27BCE), the population of the Roman empire would explode to between 55 and 64 million individuals. The city of Rome alone had a population of about a million. Just eight thousand years earlier, barely fifty humans could live together without killing each other.

With these larger civilizations came larger-scale wars. Small skirmishes between rivals gave way to organized warfare and occupation: absorbing others' territory and subjugating the vanquished. From stone missiles to nuclear missiles, from the medieval torture chamber to the gas chamber, the conceptual revolution impacted everything from art to the art of war, and the changes it wrought began to snowball.

***

A person born around 1900 in America or Europe would have entered a world in which the main engine of human transportation was the horse. The world population had grown to a staggering 1.6 billion. New York City alone boasted a population of 3.5 million human beings, with no sewage system and no flush toilets. The contents of chamber pots were tossed out of windows. There were over 150,000 horses pulling all the cabs, buses, and carts around the city. They generated an estimated 2.5 million pounds of horse manure each day, which piled up on the side of the street like snowdrifts. The situation was the same in London, Paris, and any other major urban center of the West. Many of the African Americans alive at the time remembered a life of slavery. Children under twelve were commonly employed in coal mines and factories; workplace mutilations and fatalities were common. Women couldn't vote, and were usually married before their twenty-second birthday. They would have three or four children, at least one of whom would die in its first year. Nearly one in a hundred would succumb to infection or other complication during childbirth.

It was just 150 years ago that a Hungarian physician named Ignaz Semmelweis had the insight that if physicians washed their hands between procedures, it might reduce what had become a very high mortality rate during childbirth. He was right, but because germ theory didn't exist and doctors couldn't understand why they should have to wash their hands, his discovery was forgotten for another fifty years. He died in an insane asylum, shunned by the medical community.

Until the twentieth century, humanity had almost nothing in the way of effective medicine. In the last hundred years, the power of exponential growth resulted in a nuclear explosion of medical knowledge. The American Board of Medical Specialties currently certifies more than 145 specialties and subspecialties for doctors alone. The American Medical Association's list of basic medical procedures numbers around eight thousand. Drugs.com contains references for more than twenty-four thousand prescription medications. Life expectancy has doubled since 1850.

***

In 1953, the U.S. Air Force was facing a problem. It had the funding to design the next generation of aircraft, but wasn't sure where it should be setting its sights. It had been only fifty years since the Wright brothers' rickety homebuilt airplane had hit a top speed of about 4 miles per hour at Kitty Hawk. By 1953 the sound barrier had already been broken. The Air Force brass felt that aviation technology was changing so rapidly, they needed a way to predict what was coming next. So they took the history of transportation in America, starting with the Pony Express, and made a graph. On one axis they plotted speed and on the other they plotted the year that speed was achieved. Railways had replaced horses in the early 1800s, and cars had overtaken trains in the early 1900s. Then an interesting thing occurred. The speed graph began to rise exponentially, with propeller-driven aircraft giving way to jet-propelled and then rocket-propelled aircraft in just the space of a decade.

The graph showed something the Air Force couldn't believe: that rockets would reach escape velocity—the speed needed to go to the moon and other planets—not in some distant science-fiction future, but within *six years* (i.e., 1959). It wasn't time to fund new airplanes; it was time to be funding spacecraft.

This prediction turned out to be exactly on target: the Russian space probe *Luna 1* reached escape velocity (roughly 16,000 miles per hour) and sailed toward the moon on January 2, 1959. A decade later, the *Apollo 10* crew became the fastest humans in history when they reached a velocity of 24,791 miles per hour on returning from the moon.

Apollo astronauts took a famous picture of the earth, centered on the Middle East, where the first wheeled vehicles had been invented about five thousand years earlier. The chariot was the first high-tech transportation system and the army's first tank. But around 1200BCE the Phoenicians took the transportation prize with their ships, which allowed them to carry enormous cargoes swiftly and reliably, bridging cultures and widening trade.

Wind-powered ships and horse-powered carts were the staples of transportation for thousands of years until the coal-powered steam engine overtook them in the early 1800s. A steamship could cross the Atlantic in about five days (a month faster than a sailing ship), but in 1958, Pan American 114, the first commercial transatlantic flight, covered the distance from New York to Paris in eight hours and forty minutes. In August 2001 there were over seven hundred thousand total passenger flights in the US, carrying around 56 million people.

Exponential growth allowed a small group of slightly intelligent apes to end up with a high-tech, global civilization that could send people to the moon or allow massive destruction by an angry few. A perfect storm of evolutionary factors (like bipedalism), technological advancements (like toolmaking), social organization (like hunting parties), and brain growth created a positive feedback loop in human development that continues today.

In our lives technological and social change has become almost too fast to observe. Futurist Ray Kurzweil has mapped this rate of change. As he puts it, the nineteenth century saw more technological change than the nine centuries preceding it. The first twenty years of the twentieth century saw more change than the entire nineteenth century. This pattern is continuing to accelerate, which means,

"...the technological progress in the twenty-first century will be equivalent to what would require (in the linear view) on the order of two hundred centuries." The change we will experience in the twenty-first century will be the equivalent of *twenty thousand years of change* at the rate it was occurring at the time of Rome. The next hundred years will see the creation of four or five inventions with as great an impact on humanity as the invention of agriculture or writing.

The rate of biological change in evolution is astonishingly slow. The last really large evolutionary change we went through as a species occurred when we stood up on two legs 4 million years ago. The most rapid change was the 35 percent increase in the size of our frontal cortex, and that took millions of years. For life to evolve from puddles of DNA to Ignaz Semmelweis struggling to get doctors to wash their hands took 3.5 billion years.

Technology and the conceptual thinking that creates it change at a much faster rate. We went from throwing sharp rocks to launching Saturn V moon rockets in just fifty thousand years. Ideas grow, change, compete, and become more complex, just as life forms do, through a process of artificial selection. Nature selected the genes that led to the conceptual mind capable of imagining and exchanging conceptual information with other minds. The immensely rapid growth of conceptual complexity can be thought of as the result of the reproduction and evolution of memes.

Once Descartes, Bacon, and others began formulating the scientific method for testing and proving hypotheses, the meme caught on like wildfire and thinkers everywhere developed and refined it. Standing on the precipice of a revolution in biotechnology in which we will alter the physical structure of organisms—shaping biology to match our imagination—we are entering an age when memes will, in effect, be able to engineer and control genes. We are on the brink of an age of truly extreme change.

What we know from Einstein is that both speed and point of view cause tremendous distortion in the apprehension and interpretation of reality. Near the speed of light, the effects of gravity expand; time slows down and space warps. Our perception of events is programmed by nature to assess relatively slow-moving situations quickly, to spot far-off danger on an open savanna, to read the face of an inarticulate infant, to size up a stranger encountered in the jungle. In that reality, perspective appears fixed, with us at the immovable center. In this new reality of rapid change, nothing is fixed and all is relative. There is no truth but consensus.

***

Barry Jennings was deputy director and emergency coordinator for Emergency Services Department, New York City Housing Authority. A big heavyset African American, Jennings had arrived the morning of

September 11 to work in 7 World Trade Center, a building across the street from the Twin Towers. According to Jennings, a bomb went off in 7 WTC and he hurried down to the lobby. Scrambling around the building with Michael Hess, the city's corporation counsel, they went to an Office of Emergency Management command center on another floor. "...[W]e noticed that everybody was gone ... only me and Mr. Hess were up there. After I called several individuals, one individual told me to leave and leave right away. Mr. Hess came running back in and said, 'We're the only ones up here, we gotta get out of here.'" Struggling to maneuver in the dark and heat, Jennings broke a window with a fire extinguisher.

> *Once I broke out the windows I could see outside below me. I saw police cars on fire, buses on fire.... I was trapped in there for several hours.... The firefighters came. I was going to come down on the fire hose, because I didn't want to stay there because it was too hot; they came to the window and started yelling "do not do that, it won't hold you." And then they ran away. I didn't know what was going on. That's when the first tower fell.... And then I saw them come back.... with more concern on their faces. And then they ran away again. The second tower fell.*
>
> *So as they turned and ran the second time, the guy said "We'll be back for you." And they did come back, this time they came back with ten firefighters.... All this time, I'm hearing all kinds*

of explosions. I'm thinking that maybe it's the police cars [and] buses that are on fire. I don't see ... you know, but I'm still hearing all these explosions.

When they finally got to us, and they took us down to what they called the lobby, because when I asked them, I said "Where are we?" He said, "This was the lobby." And I said, "You got to be kidding me." Total ruins. Keep in mind, when I came in there, the lobby had nice escalators—it was a huge lobby. And for me to see what I saw was unbelievable. And the firefighter who took us down kept saying, "Do not look down." I kept saying, "Why?" We were stepping over people. And you know when you can feel when you are stepping over people.

Jennings escaped with his life, and was interviewed on ABC 7 live television right on the street. His story, like that of so many other 9/11 survivors, is one of panic, confusion, and eventual rescue. But there is also one major difference: Jennings's story doesn't jibe with the official version. Seven World Trade Center was forty-seven stories tall and its tenants included several large financial firms and banks, as well as offices of the Securities and Exchange Commission, the IRS, the Department of Defense, the Secret Service, and the CIA. The official report of the 9/11 Commission states that 7 WTC collapsed as a result of damage to its structure after the collapse of the Twin Towers, made worse by several fires. There were no internal explosions in the building, especially not the multiple detonations that Jennings reported. When it was suggested that he may have heard a boiler bursting, Jennings replied, "I know what I heard. I heard explosions. The explanation I got was it was the fueloil tank. I'm an old boiler guy. If it was a fuel-oil tank, it would have been one side of the building."

Seven WTC received very little damage compared to other buildings that remained standing, yet it crumpled at near freefall speeds, as if its skeleton had vanished. The Twin Towers also fell very quickly, straight down. Many people have remarked how unusual it was that these tall, metal-framed skyscrapers could be just dropped to the ground by a fire. Supposedly they are the only such buildings in history to collapse from fir.

This sort of speculation is taken to suggest that the World Trade Center buildings did not collapse because of burning jet fuel or impact damage, but instead were brought down with explosives in a secret, controlled demolition. According to this theory, the Bush administration was complicit in the 9/11 attacks and used them to stampede a terrorized public into wars in Iraq and Afghanistan and to bargain away their freedoms for a sense of safety. As the movie *Loose Change* would have it:

> *[9/11] was an American coup; a violent and aggressive seizure of power and a transformation of both foreign and domestic policy. The slaughter*

of civilians in broad daylight as the means of promoting an agenda.

One fact adds an ominous note: Jennings mysteriously died just a few days before the National Institute of Standards and Technology released its report on the collapse of 7 WTC, which contradicted his statements.

How can these anomalies have been ignored or disregarded by the mainstream media, law enforcement, and the government? Many see this silence as a deliberate attempt to sweep these questions under the rug, reinforcing the sense of conspiracy. These apparent facts are difficult to explain and when taken all together force many people to question our understanding of what happened on 9/11, who was behind the attacks, and what the motives of the real attackers were.

Once you open the door to questions like these, the uncertainty just multiplies. All the hijackers died, but a few of their passports were found, miraculously undamaged, on the ground after the attacks. How could they have survived the crash and fires intact? Could they have been planted by agents to frame the hijackers? We were never able to question Osama bin Laden before he was killed. How do we know he was even involved?

If you already believe that the government is the enemy, or that large, powerful military and corporate entities are conspiring to control the world, that will predispose your mind to believing that the 9/11 attacks were the result of a nefarious conspiracy from within the United States: an effect called confirmation bias. Add in negativity bias—the brain's exaggeration of danger—the speed of change, and the enormity of trauma, and our old wiring makes it makes it even easier to go there.

Another cognitive distortion is called selective perception, in which we filter out information that doesn't fit into our preexisting belief structure and focus only on the information that supports our beliefs. In one experiment, viewers from two universities were shown a movie of a rough Princeton-Dartmouth football game. Although both teams committed infractions, Princeton viewers reported seeing twice as many fouls by Dartmouth players as did viewers from Dartmouth. We see what we want to see.

Human beings have dozens of known cognitive biases. These are ways in which our brains distort or misapprehend information based on the evolution of our mental architecture or brain software. With the *endowment effect,* we believe things we own are worth more than others think they are. At work and in relationships, the *self-serving bias* makes us take credit for success but deny responsibility for failure. Because of the *superiority bias,* we feel that our own positive traits are much above average, when of course most must be average. The *Semmelweis effect* causes us to automatically reject any new theory that contradicts entrenched ideas. *Normalcy bias* makes it impossible to plan for extreme situations such as planes crashing into the Twin Towers, a category 5 hurricane hitting New Orleans, or a 9.1 quake and giant tsunami in Japan. There is even a *bias blind spot,* which makes us believe that we are less biased than other people.

Just like negativity bias, all of these cognitive biases have good reasons to exist. At one time or another they were adaptive for the species. They are shortcuts for figuring out who's a friend and who's an enemy, how to avoid danger, and quick planning. Thinking uses up time and it takes energy, yet we need to think. The brain is the most energy-hungry organ in the body, consuming a full 20 percent of all calories we intake. If you require more energy, you require more food. If you waste time calculating, you might not react fast enough to danger. If you can reduce the time and energy it takes to think by substituting simple ideas for complex ones, you win.

In life we often have to make decisions with insufficient information. This was especially true in the distant past, when humans had almost no understanding of the world around them. We didn't always have all the facts before having to decide quickly and repeatedly. Biases evolved as a way of making decisions—which were just as important to our ancestors as our decisions are to us—under such limitations.

Cognitive biases are often compared to optical illusions, and the biggest cognitive bias we have is the ego; the optical illusion of our thoughts and feelings that makes us experience ourselves as separate, alone, and at the center of the universe. Our mental self-representation is itself a cognitive distortion that strikingly alters our behavior by twisting the facts of the situation. We blindly evaluate and relate to all events as if they are about us.

To have some sort of stability and consistency, we experience ourselves as a solid, never-changing self. We believe that we are something that we are not: a little mental avatar in our heads. But most fundamentally, we don't see that it is a mental representation at all. We believe it is real because it *feels* like it's real.

Philosopher and brain researcher Thomas Metzinger describes the hallucinatory nature of the ego, saying "the Ego is a transparent mental image: You, the physical person as a whole, look right through it. You do not *see* it. But you see *with* it."

The ego (which is not a thing but a function of the brain, composed of self-referential thoughts and feelings) is the ultimate conceptual overlay, and also the ultimate cognitive distortion or bias. It is a subconscious model of ourselves and our world that, just like other biases, can cause real havoc when it's taken too far.

The brain has its ruts and its million-year-old shortcuts, the behavior programs that keep us headed toward food, sex, and a safe place in the city. Stuck in our ego tunnels, we might believe we know what God wants us to do: pass out Bibles in Harlem or bring the proud skyscrapers of the unbelievers to the ground. The domesticated human will fly in space and then devise a complex plan to club a rival over the head. Yet our modern brains are not solely made up of cognitive distortions and self-delusion. Outside our conceptual contexts, emotional feedback loops, and automatic processes there are other possibilities. Though we might sometimes feel paranoid enough to believe tall tales about what happened on 9/11, the weight of evidence, Occam's Razor, and group intelligence snap us out of it and set us straight.

***

Omar bin Laden has his father's face, but there the resemblance ends. The fourth son of Osama bin Laden has a shoulder-length black mop of braids, and his beard is clipped into a short goatee. He wears distressed black jeans, a black T-shirt with a logo, and a black leather jacket covered in zippers and snaps. You would never guess that this trendy hipster working as a contractor in Jeddah spent most of his childhood in al-Qaeda training camps of one sort or another. Always without heat or running water, he grew up hanging out with murderers and watching his pet dogs get gassed. From the age of fourteen, he lived in the same house with Ayman al-Zawahiri at his father's terrorist headquarters in Afghanistan. He seemed destined to become an al-Qaeda star.

But hearing the plans for attacks that would target civilians, he had a sudden realization that he couldn't support his father's ambitions, and he escaped his father's influence. He was in Saudi Arabia when the 9/11 attacks occurred, and during a 2003 pilgrimage to Mecca he told his friend Huthaifa Azzam, "It's craziness.... Those guys are dummies. They have destroyed everything, and for nothing. What did we get from September 11?" Although he does not condemn his father's motives, he firmly believes the man's actions are wrong. "I don't think 9/11 was right personally. I don't agree with 9/11 or with any war where only civilians are dying."

The young bin Laden has since been joined by other defectors, many of whom were central to global terror. Upon realizing that he had trained the men who perpetrated the Bali bombings that killed more than two hundred civilians, Nasir Abbas, one of the heads of the Indonesian branch of al-Qaeda, renounced his activities and began working with the police. A similar wake-up call forced Noman Benotman, head of the terrorist organization Libyan Islamic Fighting Group, to forsake jihad and publicly repudiate al-Qaeda. And in 2008 the commander of al-Qaeda in Algeria voluntarily gave himself in to police after it dawned on him that the jihad he was waging was illegitimate. Even someone as committed as al-Zawahiri's good friend and terror associate Sayyid Imam al-Sharif (aka "Dr. Fadl"), formerly in the top echelon of al-Qaeda, wrote a book rejecting the group's methods.

Sometimes the fish can become aware of the water in which it swims. Even when we are awake and moving through the world, a large portion of our behavior can be on automatic—subconscious, implicit, unnoticed—until we become conscious and realize what we are doing. It happens every morning and many times a day: we drive absentmindedly to work and only come to awareness when we notice we have arrived. Waking up in this way is a small thing for a human being, something we don't even pay attention to unless somebody points it out to us. And yet this capacity to observe our own experience, to make the implicit explicit, represents an unparalleled miracle of evolution: the ability to have conscious awareness. It gives the mind the ability to see through its own biases.

We came to a kind of collective conscious awareness with the advent of the scientific method. Its emphasis on the direct observation of nature as opposed to cognitive bias meant that within two hundred years after Newton the human animal understood more about the world than we had in all previous history. Lavoisier discovered oxygen and founded modern chemistry, Vesalius and Harvey described the circulation of blood and jump-started the study of anatomy; Leeuwenhoek made the microscope and discovered bacteria, spermatozoa, and opened the world of microbiology; Linnaeus classified all known animals, the foundation of the field of biology; Adam Smith founded economics; James Hutton's ideas about rock formations established the antiquity of the earth and the science of geology, and Charles Darwin put forth the theory of evolution. The scientific revolution forever changed how human beings relate to knowledge and the world. Because it freed us from the fetters of our naïve biases and belief systems, science quickly led to the greatest wave of innovation the world has ever seen.

Science refashioned how people decided something was true: not based on our feelings and intuitions (which are subject to biases) or based on the proclamations of an authority (which can be wrong), but instead through reproducible experiment and debate. It no longer mattered what Aristotle or the Bible said, because anybody could go out into their backyard, look into a telescope, and see for themselves that these venerable authorities were dead wrong. A new awareness was emerging to question the ego's apprehension of reality, the ego's imperative to see reality as a reflection of its own sense of self-importance.

***

In 2008 the National Institute of Standards and Technology released its analysis of the collapse of 7 World Trade Center. Three years in the making, the report details how the building was struck by falling debris from the collapse of 1 WTC, which damaged the exterior structure on its south side. This debris ignited raging fires on at least ten floors of the building. Because the Twin Towers' collapse had broken a water main, the sprinkler system inside 7 WTC was unable to reduce this inferno, which eventually engulfed entire floors. Burning out of control, the flames weakened a column in the northeast part of the building, which gave way, initiating a catastrophic chain reaction of collapsing floors that brought down the building.

The report rebuts assertions of explosives being used in a controlled demolition scenario: "Blast from the smallest charge capable of falling a single critical column would have resulted in a sound level of 130 [decibels] to 140 [decibels] at a distance of at least half a mile. There were no witness reports of such a loud noise, nor was such a noise heard on the audio tracks of video recordings of the 7 WTC collapse." That level of noise is ten times louder than standing in front of a speaker at the loudest rock concert. Thousands of people would have heard it, yet nobody did.

Before he died, Barry Jennings gave another long interview in which he repeated the story of hearing explosions, but he also cast doubts on some of his own assertions, such as that the lobby was filled with dead bodies. "I never saw dead bodies. I said it felt like I was stepping over them, but I never saw any ... Do I think that our government would do something like that to its people? No, I honestly don't believe that."

Also, the impression that the building collapsed instantly has been debunked. It is clear from footage with several firefighters and others on the scene at the time that the building was bowing outward, slowly getting ready to fall, and that the fire department pulled all personnel from the site just in time.

Seven WTC fell for the same reasons the Twin Towers fell: its structure was weakened by prolonged fires. There were no deaths in the building because the rescue personnel did their jobs bravely and competently and all systems were functioning to support them.

To believe anything but that the WTC buildings were brought down by the independent actions of al-Qaeda terrorists, you either have to pretend that the mountain of contrary information doesn't exist or postulate a conspiracy so large that it would be impossible to keep quiet. Out of thousands of witnesses to an event, a few like Barry Jennings will always report something anomalous. Stress, confusion, biases, misapprehension: there are many reasons a few such incongruous reports may appear. And yet in the case of "truthers"—people who believe that 9/11 was an inside job—these anomalous reports are the only ones worthy of credence.

The ability to look at a set of possible facts and choose from among them the ones that seem to be true is an essential human trait. Even more useful is the ability of a group of people to look at the facts, debate their various merits, and arrive at a consensus of what must be true. Science required writing, mathematics, publishing, and above all a method of social interaction in which people could present their findings and attempt to reproduce the findings of others, all without resorting to violence. As a method for overcoming our biases, it is unparalleled: we no longer believe that the earth is the motionless center of the universe, that the stars are tiny campfires in the sky, or that lightning is caused by angry gods. Even though we cannot see them with our naked eyes, we know germs and atoms exist, which allows us to cure diseases and build aircraft. And because we can do testing on thousands of people, we know that our brains are hardwired for certain biases.

There is a second way to correct our cognitive distortions, though it is a biological rather than social way. It arises naturally within the individual human brain and is evolution's way of guiding us toward more accurate assessments of the world. The complexity of our brains is increasing, and with it our ability to see through the brain's self-deception, the compulsive planning, the whirlwind of feelings, and the mental prison of the conceptual mind.

***

Observe yourself reading this book right now. That is conscious awareness at work. Now wiggle a finger, and watch it wiggle. You are wiggling your finger and you are consciously aware that you are wiggling your finger. Think about how often you move your fingers without noticing that you are commanding them to move. Now picture, in your mind's eye, your mother.

Really see her face in your mind. That is conscious awareness noticing the content of a visual thought. Conscious awareness mirrors what is already present. It functions like a looking glass for our experience.

Conceptual thinking and the conceptual self occur mainly outside of conscious awareness. Most of our mental modeling of the world is done subconsciously. The underpinnings of the mental self-representation are implicit, which is what gives it such power to affect decision making and create strong biases. As the Implicit Association Tests show, our opinions about the world and our ideas about ourselves are almost totally subconscious. Conscious awareness allows us to bring these implicit ideas into the light of knowing; we can see them clearly as objects in the mind.

Examining our own experience is a very odd thing to be able to do. Probably no other animal on earth has the ability to be self-aware. Small children certainly do not, and it only comes on in fits and starts over the course of our early years. This is one reason so much of childhood is a blank: just like driving to work without remembering it, we were not consciously aware of our experience so it is not available as a conscious memory. We still lived all those years, of course, and the memories are in there somewhere, but we cannot access them because we were unable to observe them. It is related to our lack of memories of the time we spend sleeping.

Conscious awareness arises when we are able to monitor part of what's going on in real time. In the most simplified way of understanding it, conscious awareness is one set of neurons monitoring another set of neurons; the conscious brain observes parts of the subconscious brain and brings them to light. Conscious awareness can also pay attention to other things besides thoughts and feelings, of course, like external events in the world around us.

Evolution may have come up with conscious awareness as an antidote to the blind side of the conceptual mind. It can take the subconscious, implicit, biologically determined aspects of our brain—the biases, the distortions, even the ego—and expose them to the light of day. We can actually witness and assess some of the invisible backroom processes of the mind, which for every other organism, even earlier human species, would be impossible. We can open the hood and get a good look at what's going on inside us. And we can crack the code of the misunderstandings that cause us the most suffering.

***

In 1905 an unknown patent inspector in Bern, Switzerland, published five papers that revolutionized the science of physics. In the space of just six months, "a storm broke loose" in the mind of Albert Einstein, which resulted in his scientific explanation of the photoelectric effect, Brownian motion, and special relativity—which changed our notions of space and time forever—and the mathematical description of the equivalence of matter and energy in his iconic equation E=mc2. This *annus mirabilis* transformed our scientific understanding of the world and ushered in the modern era.

It also laid the groundwork for the moment in 1907 when Einstein experienced what he would refer to as the "happiest thought of my life." He had been wondering about how Newton's theory of gravity could be explained within the framework of special relativity. To do so would unify gravity and acceleration, a powerful integration of seemingly disparate phenomena under a single theoretical umbrella. As he described the moment:

> *I was sitting in a chair in the patent office at Bern when all of a sudden a thought occurred to me. "If a person falls freely he will not feel his own weight." I was startled. This simple thought made a deep impression on me. It impelled me towards a theory of gravitation.*

This happy thought led to Einstein's general theory of relativity, which underlies our modern view of the world yet has profoundly counterintuitive consequences. For example, gravity makes clocks run slower. Space is not uniform, but can be warped by massive objects, and since energy and mass are equivalent, energy can warp space too. If you travel close to the speed of light, time will almost stop for you (although you will not notice), while the rest of the universe quickly ages. These odd ramifications are just a few of the ways in which Einstein's sudden insight fundamentally undermines our naïve experience of things being fixed, permanent, or objective. It shatters our cognitive biases and forces us to look at the world in a new way.

> *A human being is a part of a whole, called by us "universe," a part limited in time and space. He experiences himself, his thoughts and feelings as something separated from the rest—a kind of optical delusion of his consciousness. This delusion is a kind of prison for us, restricting us to our personal desires and to affection for a few persons nearest to us.*

Einstein's ability to talk about the "optical delusion of his consciousness" points toward a clear view of that delusion, the typical perspective engendered by conscious awareness of the inner workings of the mind.

While Einstein is virtually in a class by himself in terms of science, he is just one of many brilliant evolutionary outliers who may have experienced expanded conscious awareness in recent human history. The very earliest writings we have that unambiguously describe this perspective are no more than twenty-five hundred years old, at the beginning of the Axial Age, when men like Siddhartha Gautama and the seers of the Upanishads described such a viewpoint. Although embedded in a matrix of mythological material, the description of the illusory nature of the self-representation, composed of thoughts and feelings, is unmistakable.

PART THREE

The Enlightenment Revolution

chapter 10

the organism is in charge

On September 8, 2001, the telephone rang at Warrick's Rent-a-Car in Pompano Beach, Florida. Warrick's was a small company tucked away in a strip mall, with just a few cars in their fleet. As proprietor Brad Warrick described his business, "A lot of criminals come here because we're a little guy, out of the way ... We don't have software in our computer system that checks the background of drivers like the major companies do."

When Warrick answered the phone, he spoke to a customer who had rented a white Ford Escort, the cheapest car on the lot, a few days earlier. The customer was a polite Middle Eastern man in his thirties who "...had very little accent and acted like he was a businessman," Warrick later remembered. "He always had a briefcase and books in the trunk." Good-looking, with wide-set eyes, a square jaw, and a large smile, the man's name was Mohamed Atta.

Atta had called Warrick to report that the Escort's oil light had blinked on. Warrick thanked Atta for letting him know that the car needed servicing. When Atta returned the car on September 9, he again took the time to remind Warrick about the oil light. The only thing unusual about Atta, Warrick said, was that "he was nice enough to let me know that the car needed an oil change." Most customers wouldn't have bothered.

This "polite businessman," who was concerned enough about how the long-term service details of a rental car might affect his fellow humans, would two days later intentionally plunge a 767 jet airliner into the North Tower of the World Trade Center, helping to kill thousands of people.

Atta not only murdered all the people in the plane he hijacked and many people in the World Trade Center, including policemen and firemen who had rushed to the scene to help; he also was the ringleader of the 9/11 terror operation that killed many more. As the head of the Hamburg terror cell at the center of the attacks, he had organized the operation, helped pick the other pilots and hijackers, and inspired them to commit an epic atrocity.

How is it that a person can worry about the minor effects a malfunctioning rental car may have on strangers yet indiscriminately slaughter thousands? Sticking with the Darwinian view that we are all human beings who evolved to live with other human beings, what allows someone like Atta to do what he did?

Mohamed Atta did not seem like a killer to his boyhood friends in Egypt. They remember him as a shy, sensitive loner who would get emotional if someone hurt an insect. Atta's father was an attorney, and his mother was a homemaker from a wealthy family. His sisters were successful: one became a medical doctor and the other a zoology professor.

Atta himself received an architecture degree from Cairo University, but his grades kept him out of the graduate program. This upset his father, who insisted that Atta study abroad and arranged for his son to travel to Germany. Two weeks later Atta was living with a German family who ran an exchange program in Hamburg, far away from the Egyptian society in which he'd lived his whole life.

Being injected into secular Western culture was not easy for Atta. An outsider alone in Germany, the architect had difficulty adjusting. His traditional Muslim values conflicted strongly with the freedom and openness of society in a modern European nation. The unmarried daughter of his host family had a baby and Atta behaved rudely toward her, presumably because he thought her immoral. This, combined with his adherence to strict dietary rules and lack of friendliness, made the situation so awkward that the couple asked him to leave. His "complete, almost aggressive insularity," refusal to clean up, and general unsociability next upset his roommates in university housing as well.

During this time, Atta's religious views mixed with political ideas to create a belief system that was toxic and dangerous. The first signs of this arose in terms of architecture. His graduate thesis was a diatribe against the modern Western skyscrapers that loomed over the ancient Syrian city of Aleppo, which

Atta thought conflicted with the ancient Arabic styles. He felt that the tall, cold buildings belittled people's privacy and dignity. Even his family's apartment block in Cairo he criticized as "a shabby symbol of Egypt's ... shameless embrace of the West." Muslim civilization was being threatened and attacked by modern Western culture, in his view, and skyscrapers were the most obvious symbol of this onslaught.

Atta felt an emotional identification with a romantic sense of "old" Islam, as symbolized by rustic homes and simple folk. The Stone Age idea that his tribe was under attack—a hunter-gatherer program developed for a family group of about thirty people, now being applied to a religion of a billion—fired up the fear circuits in his brain. That part of our minds that the neuroscientist Joseph LeDoux calls the "emotional lizard" with its limited binary response—run away or fight back—cranked into action within Atta. He chose to attack what he perceived as literally life threatening to himself, his friends, his country, and his religion. Because of his deep identifications with these things, his brain made the leap from defending his little hunter-gatherer band from intruders to attacking what he saw as the source of the threat: America.

This is one of the shortcomings of the conceptual mind: that it can take the basic survival strategies that evolved to protect us and hijack them by imagining a symbolic or conceptual threat to be real. Imbuing a skyscraper with a deadly intent toward a religion is not that different from imagining that mountains and clouds have thoughts and feelings. When this unsupported leap of the imagination happens, we go from sticking up for our family in a tussle to rounding up and gassing millions of humans we have never met. Passing on our genes by gathering energy and protecting our group becomes genocide, terrorism, and world war. The deep structures of the brain, which evolved to move us toward energy and keep us from harm, can get so turned around by ego, imagination, and identification that people even kill themselves trying to save others from phantom dangers.

Atta was living in Germany, benefiting from its top-quality educational system. He got a good job at an architecture firm where he was well liked and had German friends. In the most direct, literal sense, he was greatly benefiting from the modern, secular West. Food, shelter, education, friends, opportunities: all of these were his in European society, yet his brain chose to ignore this reality. Just as humans believe the ego to be an actual entity that can be in danger, we can believe that something like "modernity" is an actual entity that can be the cause of that danger. Symbols and ideas become personal enemies.

The view that non-Muslim, Western modernity was a death threat to his people soon permeated the rest of Atta's life. He even left a woman he loved due to her slightly nontraditional lifestyle and joined the Al-Quds Mosque in Hamburg, known for its militant, harsh, and virulently anti-Western views. Iman al-Fizazi, the head of the mosque, preached a radical version of Sunni Islam that attracted fundamentalists, and soon Atta was meeting the like-minded individuals who became the Hamburg terror cell. The people around him now were confirming his viewpoint, a social feedback loop that strengthened his ego identification.

Emotions exist to motivate and direct behavior toward benefit or away from harm. At some point Atta's rage and fear pushed him over the edge into taking action. He and his cohorts made clandestine trips to Afghanistan, where they trained in terror techniques. Educated in technical fields and acclimated to life in non-Muslim countries, Atta and his friends were exactly what Khalid Sheik Mohammed and Osama bin Laden needed for their "planes operation." These were the men who would attend flight school and pilot the jets.

By June 2000 Atta had come to America. There he and other members of his cell trained in piloting commercial aircraft in Florida, where he also rented cars from Warrick's. And as we all know, on September 11, 2001, Atta took his hatred of a symbol—the secular West—to its horrific fruition against actual human beings by flying a 767 into a skyscraper full of people he had never met.

Atta was not knocking down a building. In his mind, he was destroying a symbol of the perceived threat to himself, his people, and his religion. Evolution bred humans to sometimes give their lives to protect the women and children who represented the future of the group and carried on the genes of the family. Yet this behavior did not evolve to be able to strongly distinguish between imaginary, symbolic threats and actual threats. Atta believed a worldwide conspiracy of Jewish bankers, controlling the media, was working with America and Israel to oppress Muslims worldwide. A building full of banks in New York, one that was a massive skyscraper to boot, represented everything Atta hated, a physical object upon which he could project all his negative concepts and beliefs. Even the people in it were just symbols to him: "Jewish bankers." And so, through a convoluted tangle of identification, emotions, protective impulses, aggressive responses, and imaginary threats—all of which are part and parcel of human neurological responses—Mohamed Atta lashed out with a misguided, bestial fury.

Such is the hold the ego-construct has on us that we will commit the most heinous crimes. Of course, we will also perform the most heroic acts as well, as the FDNY and NYPD personnel demonstrated that same day as they leaped to rescue their people. It is not that identification is some kind of tragedy. It is a natural product of evolution, which arose to allow the ego to organize the behavior of the organism around things the brain deemed important threats and opportunities. It helped humanity survive. Yet humanity is now so caught up in identification with ideas rather than reality, symbols rather than actualities, that it is creating new problems instead of solving existing ones.

The conceptual mind arose because of its ability to model solutions for energy saving and survival. It allowed us to imagine new ways to fulfill the old needs of energy and reproduction and the advance/retreat protection of the nervous system. With the conceptual mind came the conceptual self, what we might call the beginning of the personal, of taking things personally. While this gifted us with a goal-oriented drive and focus, it also became the source of unforeseen difficulties.

Just as with our implicit biases, and with the transparent viewpoint of the ego-construct, even our most cherished beliefs and motivations are driven by subconscious forces. Evolution developed the conceptual mind to solve problems, which it does very well. But it also created some new problems, like the ability to link symbols, emotions, identity, and ego together all outside the scope of conscious awareness. This lack of transparency has been the source of some of humanity's most intractable difficulties: race hatred, religious warfare, ideological slaughter. It's how we do things: as though someone else were in the driver's seat.

***

In the spring of 1964 German neurologist Hans Helmut Kornhuber and his graduate student at the

University of Frieburg, Lüder Deecke, strolled down Schlossberg Hill to eat lunch in the beautiful garden at the local *Gasthaus.* While there, they shared their frustration that the brain research being done at the time around the world focused exclusively on passive aspects of the brain. They wanted to research something active; the brain actually *doing something* willfully, rather than just passively reacting to stimuli.

They decided to look into the conditions in the brain preceding volitional acts: what the electrical activity in the cerebrum looks like just before the body takes action. Using an EEG (to measure the brain waves) and an electromylogram (to measure finger muscle movements), the two recorded the brain activity that occurred when a subject flexed a finger. Exhaustively pouring over the tapes of this experiment—even running the tapes *backward* through the equipment to try to isolate the effect—they discovered what they were looking for: a clear signal in the brain that always precedes an action. This signal they named *Bereitschaftspotential,* or BP, meaning "readiness potential." Just before any action, the BP fires in the brain like clockwork, indicating that the brain has decided to initiate action in the body. BP is the unmistakable signature of decision.

The discovery of BP led University of California neurophysiologist Benjamin Libet to conduct experiments around the *conscious awareness* of willful action. He asked subjects to move their hand whenever they wanted to and report aloud when they had decided to do so. Libet found that the readiness potential signal (BP) arose about 0.35 seconds *before* the subjects reported the decision to move.

This controversial result seems to have one simple explanation, although there have been many tortured attempts at an alternative. It is strong experimental support for the idea that *all decisions are generated subconsciously in the brain.* The sense that we, the ego, have made a conscious decision is an illusion that is created in the brain *after* the decision to act has already occurred subconsciously. The idea that you decide to act is merely another one of the brain's optical illusions. Your brain—the whole brain—makes your decisions, not your ego.

There is a long history behind the idea that choices emerge from the whole brain. The Greeks believed that conscious, rational thought actually *impeded* problem solving. Socrates wrote that inspired ideas occur when a person is "beside himself," getting the conscious ego out of the way of creative leaps. The gods themselves were thought to put ideas into people's minds, and thinking just blocked these divine revelations.

In the 1920s George Wallas, an economics professor at the London School of Economics, created a four-part theory of subconscious decision making and problem solving that has become the basis for all research in the field. In the first stage, we focus on the problem or decision to be made. Let's say a college freshman is trying to choose her major. In the second stage, she *lets go of all conscious effort* to solve the problem and moves on to other activities, like playing a sport, eating lunch, or taking a nap. Wallas maintained that during this seemingly unrelated activity, the brain continues to work on the problem subconsciously.

Waking up from her nap, the freshman suddenly realizes what she wants to do with her life. It is completely clear and final—a flash of insight that is incontrovertible. This is the "aha!" moment when the solution, which the brain has come up with implicitly, moves into conscious awareness. Because the conscious mind was unaware of the problem-solving process, the solution pops into awareness fully formed, as if by a miracle. In reality the solution is not arrived at by fortuitous miracle but by the hard work of the subconscious mind. Yet from the perspective of conscious awareness, it seems like an effortless wonder.

In Wallas's fourth stage, verification, the brain evaluates the answer consciously to see if it would actually work. Some solutions that burst into consciousness seem effective at first but upon reflection don't actually pass muster. The subconscious problem-solving mechanism is not always correct or flawless. It is simply using more of the brain.

In 2010 researchers Ron Sun and Sébastien Hélie tested Wallas's theory on human subjects. They were given a coin identification problem to solve. While solving the problem, researchers interrupted them in two different ways. In one interruption, they were asked to discuss their thinking, explicitly detailing the process. Under these conditions, subjects answered correctly 36 percent of the time. In the second type of interruption, subjects were asked to work on a separate task. As predicted by Wallas, these people answered the coin question correctly 45 percent of the time. They solved the problem better by not consciously thinking about how they did it.

Researchers Mark Jung-Beeman and John Kounious combined EEGs with fMRI to hunt down concrete traces of this unconscious problem solving in the brain. By watching the EEGs to see precisely *when* the brain changed and then using fMRI data to observe exactly *where* the changes occurred, they were able to pinpoint the prefrontal cortex and anterior cingulate cortex as the regions where the process begins. These are hubs of "top-down" thinking, meaning that during the preparatory phase, the brain is shutting out distractions and focusing on the problem. Next the brain searches through every possibility, almost all of which are wrong. During this search phase, relaxation *away* from the problem is essential. Joy Bhattacharya of the University of London has shown that there is a strong pattern of alpha waves emanating from the right hemisphere at this time, indicating deep relaxation as the brain wanders purposefully. And then—snap!—we suddenly realize we have the answer.

Jung-Beeman and Kounious found that a tiny region in the right hemisphere known as the anterior superior temporal gyrus kicked into overdrive just a second before an insight flooded the mind. At the same moment, a spike of high-frequency brain activity—gamma waves—indicated that neurons from all over the brain were instantly rewiring themselves into a new network of association: the binding of an insight into a thought that was then available to conscious awareness.

All of the problem solving happens outside the scope of conscious awareness. As Jung-Beeman puts it, "At a certain point, you just have to admit that your brain knows much more than you do." MIT neuroscientist Earl Miller demonstrated that the prefrontal cortex—which Jung-Beeman and Kounious showed is the point that kicks off the insight process—is actually pursuing the answer the entire time but does not bother to share this with the conscious mind. If the conscious mind continues attempting to slog its way to an answer, it is unlikely that any insight will arise, because the widely connected right-hemisphere neurons will not go into the relaxed, alpha-wave mode in which they make far-flung associations. By flying beneath the radar of attention, our brains are able to make many more associations much faster than they could otherwise. As Miller says, "Your consciousness is very limited in capacity and that's why your pre-frontal cortex makes all these plans without telling you about it."

***

As Einstein makes clear, humans are not separate from nature, physics, and the Big Bang, although our thoughts and feelings occlude our perception of this basic fact. Human brains operate on the principles of physics, chemistry, and biology, just like other physical processes. Choices are not made by an imaginary entity behind your eyes; they emerge from the brain as a whole. Decisions are generated subconsciously, with the most significant ones rising to conscious awareness after the fact. This process works exceedingly well without any input from the ego, from the person you think of as "me": the mental self-concept. It's not you, the ego, that decides; it's you, the organism with its large brain, that does it. The ego merely hears about it afterward and takes credit.

It feels a little unusual to consider the difference between yourself as ego and yourself as an organism; it's the difference between the mental self-concept and the actual physical human, the hardware versus the software emulation.

For effective functioning the nuances of the decision-making process are better left unseen. It would burn a lot of extra resources to generate a view of this process, and it would hobble its effectiveness. Evolution does not care if humans understand how their brains work. It just requires an individual to feel that he or she as an individual organism made the choice (and not someone else) so that all actions maintain coherence. The various regions of a brain must act as if unified in order to maintain consistent and stable behavior. Decisions are delivered to you, *fait accompli,* from the massive bio-computer of the subconscious brain. It is you, the organism, not the ego, that is in charge.

***

Mohamed Atta, Khalid Sheikh Mohammed, and Osama bin Laden had a dream to slap the secular Western powers on the nose and reignite a new era of Islamic pride and influence. Harsh fundamentalists who believed that Islam is the only true religion, they felt a deep identification with the concept of "all Muslims" and felt shame and anger at their humiliation and impoverishment. In wanting to kill the enemy, their brains reacted in a Stone Age way with a Space Age plan. Trouble is, the enemy was nowhere to be found. Soviet troops occupying a Muslim country are one thing. A skyscraper full of business people on the other side of the planet is another. High-tech resources of the postmodern era such as cell phones, the Internet, flight simulators, and ubiquitous worldwide travel made their simple desire to club somebody over the head into a Bruckheimeresque spectacle of death and destruction.

While their planning, use of language, and ability for symbolic thoughts enabled them to justify and accomplish this complex mission, it was the deeper, implicit parts of their brains—filled with unseen biases, desires, needs, fears, identifications—that made the decision to do it in the first place. The ego felt threatened, the organism reacted as if the threat were physical, and the conceptual mind filled in the details.

Richard Wajda, Chuckie Diaz, Susan Frederick, Howard Lutnick, and thousands of others had their own dreams that were killed or derailed by that event. People like Pastor Terry Jones felt a similar fear—his religion and country being threatened by aliens—and reacted by calling for the burning of Qur'ans. Steven Axelrod lashed out with fearful and angry words immediately after the attack but then recanted after careful consideration. Port Authority officer Karl Olszewski spent months digging the bodies of his coworkers out of the rubble at Ground Zero and discovered a new purpose in life: to be a soldier fighting against Muslim extremists in Afghanistan. In each of them, a similar process occurred: events were filtered through the ego, with all its biases and distortions, and the organism reacted.

The self-representation of the organism in the brain needs to think of itself as an individual, the most important person in the world. This intensifies the organism's focus on survival and adaptive behavior, which is why the ego is an evolutionary success. But it also reinforces a sense of the individual's separateness from others and concentrates attention on judging relative status and worth. Everything, including other people, is assayed for its potential benefit or threat value. And the fact that other humans are interacting this way too makes it feel even worse: nobody likes getting judged by others. Biases, knee-jerk emotional programming, and the default settings of the ego all conspire to wrap the human experience in a cloud of assessments, negative self-talk, and imaginary fears that amounts to a kind of mental prison. And it is from the distorted perspective of this jail cell that the organism subconsciously decides its destiny.

***

> *A human being is a part of a whole, called by us "universe," a part limited in time and space. He experiences himself, his thoughts and feelings as something separated from the rest—a kind of optical delusion of his consciousness. This delusion is a kind of prison for us, restricting us to our personal desires and to affection for a few persons nearest to us. Our task must be to free ourselves from this prison by widening our circle of compassion to embrace all living creatures and the whole of nature in its beauty.*

When Einstein writes of getting free of the prison, the optical delusion of consciousness, he frames it in terms of individual effort. But from the perspective of human beings as part of the whole universe, a natural result of evolution and the laws of physics, it is not so much about what each individual will do, but more about what may become of the human species in the future.

It may be that this prison break is something that evolution and natural selection—the very physical causes that gave rise to humans in the first place—will address in due time.

Looking at the trends in biology and evolution that we have seen in the last four million years, it is possible to make an educated guess at the vector of future humanity. Yet the very idea that evolution could be headed anywhere specific is a controversial one. Evolution is a process that has no goal, no endpoint, nothing it is "trying" to do. It is simply happening: the result of the laws of physics applied to self-replicating, energy-gathering chemical processes.

As a physical process, evolution does have trends that are unmistakable over the long term. It tends to develop more complex, diverse, and numerous forms of life. Sometimes it goes the other way, like when a creature with eyes loses them as it adapts to a lightless cave. But these devolutions are comparatively rare. The general trend has been upward.

Many biologists make the case that human evolution is no longer natural; we have introduced so many manmade factors into our environment that our future development is skewed toward artificial selection. As science writer Nicholas Wade remarks, the first animal that we human beings domesticated was ourselves. And, since at least the time of *Homo erectus,* we have been creating tools that have strongly influenced our evolution. It has been quite some time since our evolution was natural in the strictest sense.

In human evolution there are three major factors operating in concert: a feedback loop of toolmaking, brain size, and social organization. Whenever any one of these has increased in complexity, the other two have also increased. A better tool (a spear instead of a rock) led to more food (i.e., greater energy intake), which led to bigger brains, which allowed us to organize our social structures along egalitarian lines, which meant that we could hunt more effectively and also allowed us to imagine better tools ... and so on in a self-reinforcing whirlpool of cause and effect. Which one of these factors—tools, brains, societies—is the most important is anybody's guess. Each one was crucial in human evolution, but all three have worked together over the past four million years to create modern human beings.

The most recent stage in the expansion of brain complexity has been the emergence of conscious awareness. Conscious awareness could also be called "self-reflective awareness": the ability to observe our own internal experience in real time. Previous to a few thousand years ago, conscious awareness seems to have been in short supply. There are no obvious signs of its existence, such as writings, artwork, or religious objects from that time that point to such an awareness.

With the rise of the Axial religions, we see for the first time some mention of this sort of awareness, although often set within a mythological framework. As evolution continues and human brain complexity, social complexity, and technological complexity are all skyrocketing, it raises the possibility that conscious awareness represents the upcoming change in brain function, like the arising of the conceptual mind. We could be on the brink of a new evolutionary jump.

***

Up until the 1980s, intelligence was thought to be genetically determined. You were born with a certain cognitive potential and that was that. The trouble with this view is that it carried strong racist implications: black Americans typically scored lower on IQ tests than white Americans. This did not sit well with Professor James Flynn. "It was now being argued by perfectly respectable social scientists that the IQ gap between black and white was due to genetic differences," said Flynn. "I mean it was one thing to say that Mozart had better genes for composing music than I do and it was another thing to say that on average whites had better genes for intelligence than blacks."

Flynn mounted the most extensive study of IQ test data ever done, using information from fourteen countries covering more than forty years of tests. What he found was shocking even to him. There was indeed a massive gap in IQ—15 points, which is extremely large—but it was not based on race; it was based on *generation.* As Flynn put it, "Kids who couldn't beat the average standard of today were beating quite significantly the average standard of thirty years ago. The implication was that the standard had risen and each generation of children was doing better on IQ tests." The Flynn Effect, as it is named, laid to rest racist theories of intelligence. Intelligence is increasing because of environmental factors, not genetic ones. And it seems that people on the whole may be getting much smarter, although it's happening too slowly to notice.

If you scored the tests from 1932 on today's scale, the average IQ back then would be only 80! Seventy indicates mental retardation. The Flynn Effect has been found cross-culturally worldwide, but the scientific community is divided about exactly which environmental factors are causing it: better education, better nutrition, a more cognitively stimulating environment, a decrease in infectious diseases, or an increased societal savvy about how to maximize scores when taking timed tests.

Human intelligence has been growing for the last four million years. From *Homo erectus* to *Homo sapiens,* our cognitive capacity has been continuously becoming more complex and steadily adding new abilities, such as conceptual thinking, complex symbolic manipulation, and now advanced conscious awareness. Society and technology continue to change in ways that are increasing our cognitive capacity.

The three elements of human evolution—tools, social organization, and brainpower—are still in play and in fact have reached a phase of runaway, exponential growth. In the last hundred years, the complexity, power, and diversity of our tools have skyrocketed. Our social organization is becoming radically more complex and diverse with the rise of the Internet, social media, transnational businesses, international finance, and new corporate structures. As populations grow, political structures complexify and cultures give birth to many subcultures, countercultures, alternative cultures, and sub-subcultures.

***

Imagine if Mohamed Atta had been able to observe his own ego function, with all its constituent parts, in conscious awareness, and not take things so personally. Imagine if he were aware of the unconscious biases that were driving his actions. Imagine if our subconscious motivations were available to us in our waking, conscious life. Human beings are aware of having emotions because of the bodily expression of them: butterflies in the stomach, heart racing, brow furrowing, mouth smiling. The nervous system generates a good jolt to get our attention. Because we are so lost in the conceptual mind and the stories it creates, we tend to notice these feelings only when they are already relatively extreme. When we become consciously aware of body sensation, it is possible to notice emotions as they are just beginning to arise. Atta may have been able to sense his discomfort with Western culture much earlier, when it was still barely stirring within him. He could have observed himself on the verge of taking it personally.

Atta would have felt in his body how ultra-intense the emotional expression in his body became when he heard the words of hatred against the West his *imam* preached. He would have heard how his self-talk on the topic caused the imam's words to be repeated in his mind and kept these destructive feelings simmering inside him. The brain's network of associations is constantly rewiring itself, and the neurological principle states: neurons that fire together, wire together. These words, images, and feelings began to coalesce into a tightly bound constellation of identity, an ego-construct. And most important, Atta may have noticed as this constellation of thoughts and feelings—enraged emotion, fear of annihilation, desire to protect his group, self-talk in the form of words of the Qur'an, words of his imam, and images of skyscrapers threatening his people—began to trigger more and more often as the network of associations grew.

As with most people, these associated thoughts and feelings came together in Atta with very little conscious awareness. The process occurs under the radar of attention all the time. Yet when it is exposed to the light of conscious awareness, the very act of observing it changes the process profoundly. The "prison of thoughts and feelings" cannot fuse together into a tightly welded cage from which there is no escape: in Atta's case, escape from a suicide mission.

Under the internal gaze of conscious awareness, Atta would have been able to make the unconscious process of identity construction conscious. He would have seen it for what it was: just another mental self-representation, an image in his head connected with some feelings in his body. With this sort of self-knowledge, all the energy drains out of identification. In Atta's case, it may have lost its force, no longer hidden beneath the shroud of the subconscious, to motivate him to do harm.

Humans need the ego-construct in order to function properly in the world, but that does not mean we have to live in a prison of thoughts and feelings. When conscious awareness observes the egofunction, it decouples the hidden connections in the neural network of associations that make it seem so solid and real. Instead it becomes just a formal way of interacting with the world, almost like a suit and tie that a banker puts on to go to work, with no inherent reality of its own. It is the mental stand-in for the real actor, the organism itself.

***

When the brain develops a high level of conscious awareness, it naturally begins to notice the ego-function: the thoughts and feelings, the memories and fantasies, the body sensations, emotions, and identifications that together make up the idea of "me." We know of a few extreme outliers in the past, people whose level of conscious awareness led them to see the ego as a function. One example is Siddhartha Gautama, who spoke and wrote about this experience very clearly:

> *[Joyful and happy] is one who has recognized that there is no ego, that the cause of all his troubles, cares, and vanities is a mirage, a shadow, and dream.*

As human evolution continues, and the average brain develops a stronger sense of conscious awareness, we may all begin to witness spontaneously the "prison of thoughts and feelings"—the ego—less as a thing and more as a natural function that does not wield the weight and intensity, the drama and stress, of a solid identity. The progress of evolution may create the conditions in which we no longer take things personally, from the perspective of an ego-entity, and instead notice the impersonal nature of our own experience. In some spiritual traditions, this fundamental insight is called "enlightenment" or "liberation" from the ego. Just as the previous increase in human cognitive complexity is called the conceptual revolution, we can call this upcoming change in the human species "the enlightenment revolution."

chapter 11

developing enlightenment

> *I went one evening into a dressing-room in the twilight to procure some article that was there; when suddenly there fell upon me without warning, just as if it came out of the darkness, a horrible fear of my own existence. Simultaneously there arose in my mind the image of an epileptic patient whom I had seen in the asylum, a black-haired youth with greenish skin, entirely idiotic, who used to sit all day on one of the benches, or rather shelves against the wall, with his knees drawn up against his chin. That shape am I, I felt, potentially. I became a mass of quivering fear. After this the universe was changed for me altogether. I awoke morning after morning with a horrible dread at the pit of my stomach, and with a sense of the insecurity of life that I never knew before, and that I have never felt since.*

William James had this experience at twenty-eight. Although he had just received his medical degree, he felt disinclined to practice and instead was spending his days reading every book on psychology—a brand new science at the time, in vogue in France and Germany—he could lay his hands on.

This episode of extreme anxiety, which incapacitated James for months, followed reading the mechanistic theories of the German psychologists. James was terrified by the idea that all his thoughts and actions might be just the predetermined interactions of chemistry and physics. The thought that he might be as helpless to determine his life as the epileptic patient he had seen paralyzed him.

This is how the human organism appears to the ego: a zombie of flesh and blood that is unsettling on a deep emotional level, an automatic biological machine. Despite the horror-movie nature of the vision, this episode probably represents the first step in James's journey to seeing through the ego with conscious awareness.

The proof of this assertion comes much later in James's life, in his groundbreaking classic, *The Principles of Psychology.* Here James writes at length about the self, examining this idea from philosophical, psychological, and physiological angles. He concludes that the self, or ego, is not a solid entity but is made up of parts, such as the body, the social self, and subjective impressions, and that there is a conscious awareness—a skill, a perspective, not an entity—witnessing all of these. James saw with great clarity that the ego is a function that coordinates all parts of the self, not a thing, and that this fact was noticeable through conscious awareness. In fact he created an entire psychology based on his understanding of this.

The terror he experienced during his insight into the illusory nature of the ego, the underlying automatic nature of the brain, and his revulsion at the organism itself are fairly common reactions to the first experience of this insight.

The Buddha commented that even a doubt about the solidity of the ego is enough to tear your perception of reality to shreds. It's an unsettling insight, to say the least, but it's an inevitable one. Serving the needs of the ego has been at the root of the mistaken perceptions that have gotten us into the difficulty we're in today. Its cognitive biases and emotional identifications are the start of all wars, the cause of unnecessary violence; and because it specializes in taking things personally, it's pretty much the cause of every depressing moment we've ever had.

Conscious awareness sees through the fiction of the solid self and so the biases disappear. It's a capacity that shows up in our own development and we can recognize its appearance at every stage, especially when we get a glimpse of the nonpersonal nature of life.

***

We're born in a remarkably open and helpless state and must go through a long and difficult process of development and education in order to become a fully functional adult. This begins with physical development and includes learning to see, to walk, to eat, and to recognize our parents, siblings, and the objects in the world around us. Learning to call a dog a dog, a tree a tree, and a house a house takes years. Conscious awareness shows up in fits and starts when we begin to have memories. It's the first evidence that we are capable of observing ourselves.

Although aspects of our personality and disposition are genetically determined, our social environment programs our ego, identity, and belief systems on many levels. A baby born in America is likely to grow up speaking English, believing in Jesus, the Bible, and the Constitution. The same baby born in Saudi Arabia will speak Arabic, believe in Allah, the Qur'an, and the king. We think of these things as deep, personal choices—the essence of our identity—but there is often not much choice involved. It's a matter of chance. If we were born in ancient Egypt, we would speak classical Egyptian and believe in Osiris, Isis, and the pharaoh.

As our brain matures it begins to generate mental images of ourselves and the world around us. We tend not to experience a tree as a matrix of sensory impressions—shades of green, texture of bark, smells, sound of rustling leaves—but instead reduce it to the word "tree." And then "my tree in my yard." The same is true for our mental representations of physical objects as well as events, feelings, and ideas, using all of these symbols to assemble the mental structure of the ego.

Creating the foundation and subsequent layers of this *personal self* happens automatically while we are growing up. Just as our legs grow longer, our heads grow larger, and our sexual features take shape, the ego, too, emerges with its autobiography over a period of many years, a compelling narrative of unconscious drives setting up for the inevitability of every thought and action.

But then you find yourself thinking, *Why am I doing this? What's motivating me?* This is conscious awareness. The moment we break free from the trap of emotional identification—American or Muslim, Democrat or Republican, son of Osama bin Laden or child of a 9/11 victim—and examine the bigger picture of commonality is the moment we experience conscious awareness at work. When we notice that our cherished opinions are an arbitrary combination of thoughts and feelings—which could have been different or even opposite if our external circumstances were altered even slightly—we are experiencing the mirror of conscious awareness reflecting the nature of the ego. Under the light of conscious awareness, we feel the rigidity and constriction of the personal self and directly perceive the suffering inherent within the prison of the ego's unbending biases, opinions, emotional attachments, subconscious evaluations, and identifications.

Yet conscious awareness is itself not locked within the bars of this prison. It is capable of reflecting the nature of the personal self from an impersonal, non-identified perspective. And this is the key to its liberating effects. Just as the Apollo astronauts could take a picture of the earth from space, for the first time seeing the planet from outside the planet, conscious awareness allows us to see the personal self from outside the personal self. And when we are outside the prison, we are no longer confined by its locks.

***

In 1866, German zoologist Ernst Haeckel proposed the idea that a human fetus in the womb goes through all the stages of the evolution of the species. This "recapitulation theory," as it is called, caught on and became a fixture of the popular imagination, despite the fact that it is not literally true. Human fetuses don't develop actual gills, as the story would have it. The idea does hold water in a general sense, however. Structures that evolved earlier in the history of the species also develop earlier in the individual fetus. For example, the backbone forms in the beginning stages of fetal growth, and the cerebrum, which was a much later evolutionary development, emerges last. And it is true that a human fetus develops a tail, like that of our monkey ancestors, which eventually recedes into the tailbone.

Recapitulation theory can also be applied to the education and development of a child. English philosopher Herbert Spencer felt that the education of a single individual recaps the course of the history of civilization. Just as civilization invented basic writing and math first and then moved on to more complex activities such as literature and trigonometry, children must learn their three Rs before they can move on to Shakespeare and algebra.

The same is true of human psychological development after birth. As a child grows from an infant to an adult, it gradually passes through all the stages of human evolution since *Australopithecus.* Just as the first step in becoming human was to stand upright, a human infant spends almost a year crawling around on all fours before it learns to walk. We then dedicate a tremendous amount of time and effort to learning to manipulate things with our hands, just as our ancestor *Homo erectus* did. Meanwhile we are gradually getting the hang of human speech, like Neanderthals did. Developing a mental self-concept, as *Homo sapiens* did in the cultural revolution, starts at around eighteen months.

Mapping the growth from child to adult has been a vital concern of psychology and neuroscience for at least the last hundred years. Developmental theory, as it is now called, has many facets. There are models of cognitive development, models of moral development, and models of ego development. All of these and many more divide up the life of a human being into a predictable sequence of stages or phases.

This agrees with our intuitive understanding of how people grow up. For example, there is the "terrible twos," in which it seems that "No!" is the only communication. Or the difficult teenage years during which young people turn away from their parents, sulk, and struggle with identity. These things just seem to go in phases that start suddenly, gear up and peak for a while, and then die away almost as quickly. Humans are experts at pattern recognition, which is why grandparents can advise their children, "Oh, it's just a phase. They'll get over it." They have seen enough children grow up to recognize the template of development.

Various researchers, mainly psychologists, have taken our intuitive knowledge of such phases and teased out the scientific background and understanding for them. They have laid them out in sequences that rigidly delineate the stages and substages taking the ego as the equivalent of the self. While real-life stages and phases are never as clear-cut as the systems would make them out to be, these developmental models give us real insight into the similarities of the human experience and have become the basis of systems of education, psychology, interventions, and treatment modalities.

We found, however, that no developmental model directly addressed the potential for human development from an evolutionary perspective. While each of the existing models contributed an important piece of the puzzle, none questioned the possibility of the ego being an adaptive tool, a conceptual representation coordinating the human being. In addition, there was no adequate model that included development beyond ordinary adulthood, beyond the ego, the "prison of thoughts and feelings." The more we investigated the nature of the ego, the more we saw the possibility for an expanded model, which we call the Phase Model of Development.

The diagram below illustrates the core of the model.

This model proposes three major phases of human development: the *pre-personal phase,* the *personal phase,* and the *post-personal phase.* These three phases are defined by the individual's primary orientation in life. The pre-personal phase denotes a physical orientation, the personal phase a conceptual orientation, and the post-personal phase is when an orientation toward conscious awareness becomes primary.

The Phase Model of Development Graph

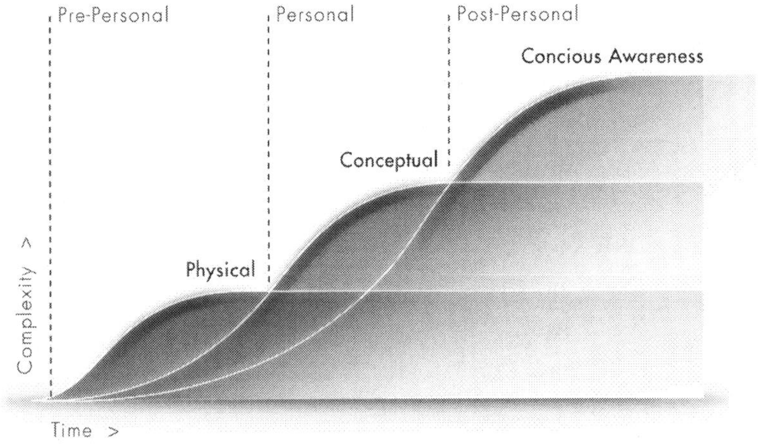

Each of the three curves in the diagram illustrates the arising, expansion, and leveling off of one of three distinct capacities: physical, conceptual, and conscious awareness that allow for the three phases—pre-personal, personal, and post-personal—to develop.

For example, when an individual is born, the physical capacities are barely functioning. The line ramps up sharply as the body and senses come fully online. After about eighteen years of age, this line levels off and stays pretty much the same for the rest of life. The second line denotes the development of the capacity to conceptualize. When it becomes primary, the pre-personal phase ends and the personal phase begins. The third line depicts the capacity for conscious awareness, which, when it becomes sufficiently developed, marks the beginning of the post-personal phase. In each phase, the other two nonprimary orientations are present but to a lesser degree than the primary orientation.

The shift from a personal to a post-personal orientation can be termed *the process of enlightenment.*

The Pre-Personal Phase

The pre-personal phase begins at birth. It is a purely physical orientation, based on genetics and environmental conditions. The body is just beginning to grow, and the senses are not fully online. A newborn can only see light and motion. It takes about six months before we can make out details, and three or four years before we have 20/20 vision. Most animals are born able to stand up and walk, but it takes months before a human infant turns over by itself, crawls, and begins to walk. Play and interaction with others boost the development of physical and sensory capacities. The body grows very quickly: an infant's weight will double in five months, triple in twelve months, and quadruple by twenty-four months.

An infant can experience the basic emotions—anger, joy, sadness, disgust, fear, and surprise—at birth. These emotions arise in the body, the nervous system's reaction to the basic needs of eating, sleeping, getting rid of waste, and proximity to those who care for its basic needs. A newborn cries when hungry but is not conscious of the fact that it's crying or hungry. A newborn cries when left alone for a long time but is not conscious of the fear it is expressing. It doesn't have the brain complexity to be consciously aware of the sensation of hunger, or to form a concept of being hungry, or the sense of fear and the concept of abandonment. Behavior at this stage is based on innate adaptive reflexes, with learned behaviors slowly accruing.

An infant's brain develops very rapidly, multiplying the number of synapses by a factor of twenty in just the first three months. The brain triples in weight, from thirteen to thirty-eight ounces, in the first three years. This increase in weight is the result of the buildup of the fatty substance myelin forming around the neurons, enhancing their conductivity, and also the increase in dendrite growth—creating synapses through learning. As the biological computer boots up, the capacity to form concepts slowly emerges and the baby picks up some basic ideas, such as "Mommy" and "Daddy," fairly quickly. Use of language indicates the progress of the capacity to conceptualize (although the two are not the same thing). As the child uses more words, more concepts are formed in its brain.

Sometime in the first two or three years, the capacity for conscious awareness also begins to appear. It is possible that memories of early childhood indicate the onset of conscious awareness. This may be why our early memories are so spotty; conscious awareness was likely intermittent at first, just like conceptual awareness.

During the pre-personal phase the capacity for both conceptual thought and conscious awareness continue to develop, albeit at a leisurely pace. With the passing years the signs of the growth of these abilities, especially the increase in conceptual thinking, become obvious. Simple sentences turn into more complex ones, and a child can anticipate birthdays and holidays as its sense of time develops.

During these first years of life, environmental conditions are crucial. On top of the basics of physical safety, food, and shelter, a sense of belonging is essential for all subsequent stages of development. A fragile sense of belonging can leave the individual prone to lifelong anxiety.

The primary orientation of childhood is physical: children become increasingly aware of what they want and can verbalize it. Instead of crying out of hunger, now they can say, "I want cookies and milk!" Concepts of real objects, fantasy, and magic all blend together. Santa Claus, being a beautiful princess, and Luke Skywalker are just as real to them as home and family. Their symbolic model of the world has not yet distinguished the real from the imaginary.

The ego makes its first appearance at around eighteen months. As the years progress and the capacity for conceptual thinking and conscious awareness increase, a mental self-concept begins to crystallize. As the child becomes increasingly self-aware, social emotions such as embarrassment, shame, guilt, and pride show up and with these a more differentiated ego develops. Identification with the physical body and physical possessions, which is the hallmark of the pre-personal phase, shifts toward identification with a personal self.

The Personal Phase

The personal phase is marked by the expansion from a mainly physical to an increasingly conceptual orientation. The capacity to form mental concepts was of great value to our ancestors. Generating a mental model of our physical environment allowed us to manipulate our environment in sophisticated ways. We could observe how things worked and imagine how they might work better. We could outsmart our natural enemies, store food for droughts or winters, and greatly improve our survival chances.

During the personal phase, the physical orientation doesn't disappear. It is of course still necessary to have awareness of our bodies and the physical world. However, the conceptual orientation of the personal phase means that the individual relates to the environment *primarily* through concepts. Sensations are not simply experienced but are conceptualized and evaluated. The flavor of chocolate doesn't only get experienced; it is analyzed and subconsciously compared with previous experiences of chocolate, other foods, and other experiences. In addition, the brain maps where the chocolate came from, how to get more, and how to avoid losing the chocolate we already have.

The brain maps and symbolizes every aspect and potentiality of the physical environment. "I'm worried that my computer password is too weak." "Gotta make that meeting in order to get the contract." This type of mental experience overshadows the sensory experience it is evaluating.

The radical expansion into the mental realm of concepts, beliefs, and imagination happens in adolescence and early adulthood, but it doesn't happen suddenly. We all know how a teenager vacillates between being a child, lost in physical sensations and emotions, and becoming an adult with the capacity to take a mental perspective on physical phenomena. It's a dynamic process. Eventually the ability to conceptualize, to think about everything, becomes central and completely natural. Events are related to personal self and appraised from a personal perspective. The ego is tremendously bolstered at this time:

the teenager takes everything personally and needs to be the best at whatever he or she is doing.

With the shift from a physical to a conceptual orientation, identification shifts from the body to the concept of the body. We evaluate the physical body by comparing it to other bodies or an ideal mental image of a body. Conceptual orientation includes evaluative aspects. "You have a great body," "My hair is ugly," "She is too fat." Comparison, ranking, and judgment become central preoccupations.

This is especially true in the social realm. Our ego rises and falls according to our standing in the group. The longing for acceptance and the fear of rejection influence every thought and behavior. We want to look good and avoid looking bad, to fit in and to succeed. For social animals, being part of a group is essential for survival; high standing in the group will bring greater resources and better mating opportunities. It's no different with human beings. In order to succeed, we strive to be an asset to the group and to do whatever we can to look sexy and be well liked, connected, wealthy, intelligent, powerful, and important.

We constantly monitor social feedback and adjust our behavior to maximize social status. The personal self is in charge, oblivious to the underlying drives and motivations that direct this behavior. The brain appraises things that are "good for me" or "bad for me" with little or no conscious awareness. We automatically judge other people based on our conditioned beliefs and concepts. We like them or we don't, they're nice or arrogant, stupid or smart, according to whether they threaten or bolster our standing in the community.

As the personal phase matures, the ego crystallizes into a more solid, rigidly defined form. We have a set of beliefs about what's right and wrong, desirable and not, and a large store of autobiographical history to relate to. The ego is in full bloom and seems to run the show. Social emotions such as embarrassment, guilt, shame, and pride strongly inform our behavior. We connect to those who confirm our beliefs and reject those who don't. We form mental self-concepts that turn us into increasingly distinct, increasingly separate individuals.

When we identify with a distinct mental self-concept, it triggers a painful *feeling* of separateness. We evolved as group creatures, and solitude, even mental solitude, is unpleasant. Prisons use solitary confinement as a form of punishment, and if it is continued long enough or strictly enough, it is sometimes considered a form of torture. Because of how it separates us from others, such suffering is at its worst when self-identification is strongest. For many adults in the modern world, this feeling of alienation and separation persists for a lifetime. Identification with a mental self-concept during the personal phase lies at the core of the human condition.

This is where most developmental models leave us: inside the optical illusion of the ego, trapped in the prison of thoughts and feelings. Even Freud felt that the best a person in therapy could hope for was to go from abject misery to "ordinary unhappiness." This is shortchanging the potential of the human brain and the impulse of evolution. In our opinion there is a natural next phase, a phase that leads to the possibility of an enlightened humanity.

The Post-Personal Phase

Conscious awareness begins to develop fairly early in life but usually doesn't get recognized as a distinct ability of the brain. Conscious awareness is a higher-order cognitive function that allows us to witness body states and higher conceptual processing as they are happening in the field of attention, instead of recognizing our actions and attitudes in hindsight. The capacity for conscious awareness relies on advanced levels of brain development, so it takes many years to become the dominant orientation, if it ever does. It is, essentially, the brain reflecting to itself both mental phenomena such as thoughts and physical phenomena such as sensations and emotions.

The capacity for some elderly persons to laugh at the folly of younger people taking themselves too seriously, or to laugh at themselves for the kinds of mistakes they used to make, points to the effects of conscious awareness growing more pronounced over time.

The capacity for conscious awareness is evolving because it is adaptive, much like the functions of emotions and conceptual thought were adaptive in bringing us out of a kill-or-be-killed, hand-to-mouth existence. Being consciously aware allows us as a whole person, the organism, to observe the workings of the body and mind from a more open perspective.

One of the illusions of ego is that identity and personality are solid and fixed. We now know from neuroscience that the brain has the potential for plasticity; it can drastically rewire itself. Conscious awareness enhances the ability of the brain to reorganize itself and optimize its functions. Without that ability, we would never be able to learn new skills or adapt to the changes of maturity.

Some basic behavioral and conceptual grooves are laid early in life and can be difficult to rewire, which is why we often keep making the same mistakes. For example, most self-improvement is intended to stop negative self-referential feelings, such as anger toward oneself, or to correct low self-esteem. Depending on how negative the self-image is, this approach can take decades without the guarantee of success.

If the capacity for conscious awareness is well developed and is able to experience the ego-as-function, these negative feelings have nothing to stick to. You can feel badly about doing something stupid, but the aftereffect of self-torture over a mistake won't be there. This applies particularly to feelings of embarrassment, shame, and self-loathing. There is no solid self for them to adhere to once the ego is understood as a function, and so they cause far less suffering.

Conscious awareness functions like a mirror: it reflects everything and holds on to nothing. It observes and reflects the attachments and emotions of the ego, but it does not subconsciously get stuck on content as the conceptual mind does. The mirror of conscious awareness reveals the subconscious workings of the ego, identification, and biases. Consciously noticing these features already begins to change them. Ballet dancers often use a mirror to notice the details of their movement they wouldn't otherwise see, and this improves their performance. Similarly, conscious awareness, by mirroring the normally hidden details of thought, allows enhancement of mental functioning.

The stuck places of emotional attachments, the warped and twisted perspective of biases, and the underlying black hole of ego identification that sucks every thought and feeling into its conceptual framework: all can be seen clearly in the mirror of conscious awareness. When the brain consciously notices these things and begins learning, it starts to reweave all of this faulty wiring into a much more efficient form that doesn't waste energy on needless suffering, worry, and ego-defense.

With the onset of the post-personal phase, the biggest change occurs with the concept of "me," the ego. Paradoxically, the personal self is not personal anymore but is understood as a mental concept coordinating the organism—body, mind, emotions—as a whole. This is why we call this phase post-personal: the viewpoint of conscious awareness is inherently impersonal.

The post-personal phase is how a person experiences the jailbreak that Einstein proposed: "...to free ourselves from this prison." With liberation from the prison of thoughts and feelings, the illusion of the ego becomes self-evident and the sense of experiencing ourselves, our feelings, and our thoughts as separate dissolves. It is replaced by an experience of openness, freedom, spaciousness, and liveliness—like a prisoner who's just been released from jail.

***

The transition between the personal and post-personal phase is as dynamic as the one between the pre-personal and personal. It takes many years for our physical body to mature, and it is an even longer process for our conceptual capacities to reach their peak effectiveness. In the same way, the development and stabilization of conscious awareness is a prolonged, dynamic process.

Conscious orientation allows us to experience the full range of physical sensations—the pleasure of chocolate or the pain of a twisted back, the joy of a sunrise or the grief over the loss of life—but without the constricting lens of the ego, which takes it all personally: *how can this be happening to me?* Being consciously aware is not an escape from life but the unfettered experience of it. It allows for full, intimate contact with the raw sensation inherent in being alive.

For conscious awareness to become strong, it is essential for basic needs to be met: physical safety, food, shelter, and a sense of belonging. Physical or psychological damage or pathologies present an enormous obstacle. Large holes in early development, difficulties with mental health, neglect, or cognitive deficiency may impede the development of conscious awareness.

When basic needs are met and a healthy developmental process is in place, conscious awareness can be developed like any other capacity: by using it. Children develop motor coordination by playing. We develop conceptual capacity by training in school. We can develop our physical capacities by physical movement and our conceptual capacities by engagement with complex problem solving. In a similar vein, we can develop conscious awareness with deliberate focus of attention on being consciously aware—particularly becoming aware of our thoughts and feelings. Especially useful is an environment where conscious awareness is valued and trained. Mindfulness meditation and other contemplative traditions, at their core, foster the development of conscious awareness.

The beginning of the post-personal orientation represents enlightenment as a developmental phase. Enlightenment begins here but does not end here. It is an emergent process that continues to deepen, expand, and open throughout life. Seeing through the solid-seeming ego and experiencing it as a function instead of as the boss is just the beginning. Conscious awareness is an extraordinary capacity that can bring our lives to a new level of rich and fulfilling experience, liberated from the prison of self-identification.

***

Post-personal orientation, by definition, includes conceptual and sensory awareness, but it goes beyond both. This means that conscious awareness is able to see what used to be personal as a function that is inherently *impersonal.* Our immediate, poignant, urgent concerns do not disappear but are instead seen as naturally arising phenomena of the living organism. And our emotional investment in these phenomena diminishes with time.

Because there has been so much focus in spiritual and religious circles on destroying or negating the ego, it's important to reiterate that the mental self-concept does not disappear. If it did that would constitute a pathology and lead to dysfunction—to people believing that conventional reality isn't real and they might be able to pay for groceries with pebbles.

The ego keeps the organism functioning in a coherent manner. Just as the heart will continue to beat and the lungs will continue to breathe, the brain will generate the mental self-concept and will continue to assess all events and appraise their usefulness. We need to know that our organism is separate and different from other organisms in order to eat, sleep, move, and behave in a way that supports our survival. It makes sense to save for retirement or send the kids to a good school. The ego function will not go away.

The ego orientation tends to be strongly identified with political and religious views, and individuals have intense reactions to views that other people hold. From the personal orientation, everything is evaluated as either a threat or an opportunity for the self, and religion and politics are no exception. We feel safe when others agree with us and threatened when they do not. With a personal orientation, we're emotionally dependent on whether or not other people agree with us; different belief systems threaten the biases of the ego.

Conscious awareness has historically been interpreted as something otherworldly. From an ego perspective, conscious awareness can feel threatening; it can seem as though we are taking on a perspective that has been attributed to God or an eternal soul. But the ego transcendence that comes with conscious awareness is a natural ability, one that will flourish when we understand the function of the ego.

The post-personal viewpoint reveals that political ideas and even religious beliefs are nothing more than concepts supercharged by emotion, evolutionary leftovers from the days when common interests meant survival in the wild. While the objective validity or practicality of those beliefs can be debated and analyzed, no *identification* with them is necessary. Even when these intensely personal feelings arise (and they will), they cannot find any purchase or handhold within the post-personal perspective. The ego's need to fight back over illusions is short-circuited.

Once we realize that nothing is personal and no one is in control, we stop judging others so harshly. This is not a saintly gift but a natural consequence of the recognition that no self-entity is in charge. An emotional response triggered by our automatic appraisal of others' behavior is just that: automatic. But if we're consciously aware of that reflex, it can well up and fade without us needing to take it seriously. Our Stone Age brain will keep triggering evaluative reactions to what we perceive, but the mere awareness of this allows us to refrain from overreacting.

***

Things as they are...

WALLACE STEVENS

The conceptual mind, focused and commanded by the ego, was favored by evolution because it allowed the humans who had this capacity to out-plan any humans who did not. Its job is to predict what can be, to imagine how things can be different, for better or worse. Imagination allows us to predict scenarios that increase safety and decrease danger. This has the advantage of allowing us to create art, culture, science, language, medicine, and so forth. We would not be humans without this ability, and it is useful that we have it. But on the down side, it makes it very hard for most people to accept anything as it is. We tend to struggle and contend with every situation, constantly attempting to control it and force it to conform to our ideas of how it should be.

With a conscious-awareness orientation we can accept, even welcome, everything that occurs in life. Acceptance, a sense of openness and relaxation with what is going on, becomes our first response to events and situations. This reaction is the outcome of a deep, visceral understanding that things can never be any different than they are at the moment. Any thought of how a situation could or should be different is just a concept in our minds that has nothing to do with current reality.

This doesn't mean that we condone or agree with everything that is happening. Remember, a post-conceptual orientation still includes a perfectly functional conceptual apparatus. We remain capable of seeing situations that need improvement and capable of acting on that understanding. And if we are being physically threatened, we will still react to cut off the threat. But in the post-personal orientation, an individual is *not resisting* reality. It is what it is. Every moment is the outcome of the process of the entire universe up to this point. We give up the illusion that we can manage conditions and will someday have everything under control. It's possible to know this intellectually, but it is quite different when we experience this from a post-conceptual understanding. Having a direct, embodied experience of the ego-as-function is fundamentally different from just knowing it as a concept.

In the post-personal phase we come to understand that whatever our eventual reaction to a situation will be, it will arise spontaneously from the whole organism. Lack of identification with actions creates a clear understanding that this is what bodies and minds do; they will succeed or fail, but with conscious awareness these actions are not personal. This leads to comfort with the way things are, not as some kind of intellectual stance but as the natural result of no longer being identified with concepts.

This deep acceptance doesn't mean feeling good all the time or becoming divinely serene, overlooking all pain and suffering in the world in some kind of glassy-eyed stupor. Rather it means being highly alert to but not rattled by even the most intense emotions, the most overwhelming circumstances. A post-conceptual orientation fosters the understanding that everything is always just exactly the way it is, and in that moment it cannot be any different.

This level of acceptance relieves stress. The Stone Age brain did not evolve to cope with the tidal wave of information we are subjected to in the modern world. Coping with the sheer volume of information we're trying to mentally digest requires all our processing capacity. No wonder we're exhausted, anxious, and chronically stressed out. With the acceptance of the situation that comes with the post-personal phase, our emotional reactivity decreases dramatically. Even though mental evaluation continues, it triggers our emotional reactions far less. The result is that our levels of stress, anxiety, and depression decrease dramatically. And it leaves us feeling free and open, able to respond to life.

***

What does the terrorist attack on September 11 have to do with the enlightenment revolution? We have been including 9/11 stories in this book because in this one event we can find indications of all phases of human evolution, including the coming phase. We can see the radical advances we've made in technology, from a rock to a 767 airliner. In the personal stories of survivors like Richard Wajda, Chuckie Diaz, Susan Frederick, Howard Lutnick, and Karl Olszewski, we can see how the emotional brain works, from our lizard-like fear center to our higher social emotions. In the stories of terrorists such as Khalid Sheikh Mohammed, we can see how the planning mind works, and Mohamed Atta reveals how belief systems and emotions can come together to create imminent threats out of potential threats. And the life of bin Laden makes the travesty of the ego, with its unconscious biases and identification that can spin into megalomania and homicidal grandiosity, all too apparent. People like Terry Jones, the Qur'an-burning pastor, and the truthers who think that 9/11 was an inside job demonstrate the societal contraction—a sort of cringing of the group ego into unconscious biases—that is one reaction to the attacks. Stories of bin Laden's son Omar, who quit al-Qaeda, and other former terrorists like him, show that conscious awareness of something that was previously subconscious is not only possible but can actually lead to a deep reevaluation of motives and behavior.

Remember Sonia Puopolo, one of the first people murdered on Flight 11? In 2007 her daughter Tita Puopolo became the first family member of a 9/11 victim to travel to Saudi Arabia. There she spoke at the Jeddah Economic Forum as part of her involvement with the Saudi-American Exchange Program. She said that she was attempting to foster:

> *...diplomacy efforts and people-to-people communication so that we can better understand each other, respect one another's values and views and, therefore, further peace, understanding and tolerance—tolerance of different cultures and cultural backgrounds in this particular region of the Middle East. For me in particular, my mother was very involved in philanthropy in the U.S. She emphasized how important it was for tolerance.*
>
> *I was nervous about being here as the first 9/11 family member. Things run through your mind. I myself am so open, but I think of those*

> *who are still unconscious, so unaware. They couldn't even begin to understand why women wear the abayas [long robes] and the men wear these beautiful white gowns. It is just a way of being. In life, we all have our perceptions of things. We learn to perceive things. If we shift our perceptions then we can live in a more peaceful existence.*

Tita even met with members of the bin Laden family, crashing through the mental and emotional boundaries of friend and enemy to find the commonality of the human condition.

Over a hundred years ago, William James experienced the ego as a function. At first it terrified him, but over the course of his life, his realization deepened and informed his seminal contributions to psychology. Today increasing numbers of people are beginning to touch the edges of the post-personal phase. As more individuals begin to see beyond the confines of their own cramped ego prison, society as a whole will evolve one of the most exciting shifts it has ever made: giving up the illusion of control as the most adaptive way forward.

chapter 12

the rise of an enlightened humanity

On the night of January 28, 2011, Wael Ghonim stood in a sea of demonstrators in Cairo's Tahrir Square. Following Friday prayers, tens of thousands of people—men, women, and children—had flooded out into the streets to protest police brutality and the government of President Hosni Mubarak. Things were getting crazy, with confused reports of escaped prisoners and looting. The police had disappeared from the streets, replaced by the military in great force. The sight of tanks, machine guns, and soldiers looming over the square led to fears of a massacre. As the evening progressed, pro-Mubarak demonstrators arrived, clashing with revolutionaries and causing several deaths.

In the midst of this mayhem, Ghonim was singled out by State Security Investigations Service men and whisked away. The SSI was Mubarak's secret police, infamous for mistreatment and torture of prisoners and with immunity from prosecution. They held Ghonim in secret, incommunicado detention. Nobody in the outside world had a clue of his whereabouts or condition. It was just known that he had been disappeared by the secret police: never a good sign.

Google, his employer, put out an emergency bulletin in search of news about him. Ghonim's last communication, via Twitter, was cause for concern: "Pray for #Egypt. Very worried as it seems that government is planning a war crime tomorrow against people. We are all ready to die."

While the world waited to learn Ghonim's fate, he was somewhere in an SSI prison. Blindfolded and handcuffed, he had plenty of time to think about how he had ended up there.

Ghonim, an Egyptian, worked as Google's head of marketing for the Middle East and North Africa and lived in the United Arab Emirates. He had been running a Facebook page, "We are all Khaled Said," memorializing a young businessman brutally beaten to death by police outside an Alexandria cybercafé. Rage at Said's killing, fanned by the morgue photographs of his disfigured face, snowballed to the point where the page had more than four hundred thousand followers venting their anger and dissent. Feeling the growing mass of discontent, Ghonim called for protests on National Police Day, to be called the "Day of Wrath." This call was echoed over Twitter and YouTube, and reinforced by many others.

Hundreds of thousands of people all over Egypt poured into the streets, bringing the country to a standstill, a response that exceeded Ghonim's wildest dreams. He flew from UAE to take part in the Cairo demonstrations, where he was swept up in a few heady, unreal days of upheaval among the crowds in the open air. Then, just as suddenly, he was alone, blindfolded and handcuffed, undergoing interrogation in a prison. The investigators relentlessly asked: Who was behind the revolution? Which forces were controlling the Internet groups, the Facebook page? They accused Ghonim of betraying his country to foreigners who were trying to bring it down.

After twelve days of questioning, Ghonim was released to an Egypt he had never seen before. While he was in prison the protests had grown larger. He was whisked to a television studio of the privately owned DreamTV, where he gave a shockingly emotional interview. When host Mona el-Shazly showed him pictures of young people who had died in the protests, Ghonim broke down in tears, saying that it wasn't the protesters' fault this had happened; it was the fault of "every one of those in power who doesn't want to let go of it." He became so emotional that he fled the studio in the middle of the interview.

This appearance made Ghonim a national hero and galvanized the resolve of the Egyptian people. But he steadfastly denied any such heroic role. Instead he emphasized the egalitarian nature of the uprising and the role of social media, calling it Revolution 2.0. Speaking on *60 Minutes,* he said:

> *Our revolution is like Wikipedia, okay? Everyone is contributing content, [but] you don't know the names of the people contributing the content. This is exactly what happened. Revolution 2.0 in Egypt was exactly the same. Everyone contributing*

> *small pieces, bits and pieces. We drew this whole picture of a revolution. And no one is the hero in that picture.*

Revolution 2.0 used Twitter, YouTube, cell phones, e-mail, and text messaging to enable the Egyptian demonstrators to organize, motivate, and support themselves. Technological changes created new methods of social organization, which were instrumental in overturning a fossilized dominance hierarchy in that country, from which one of the 9/11 hijackers had come.

Officials were savvy enough to attempt to block Egypt's Internet access in hopes of jamming the opposition's signals, but in the decentralized, networked world of modern communications, there were too many alternate channels. The old systems of information control and government oppression broke down completely.

As of this writing, several countries in the Middle East and Africa are in revolt or undergoing profound changes in government. Hosni Mubarak has been arrested and is facing trial and the SSI has been officially disbanded, but the fate of the new Egypt is unclear. The situation is in flux, and no one knows which way things will go or how they will turn out. It could be the rise of true democracy in the region, or these fledgling movements could disappear under the cowl of some new oppression. But one thing is certain: the human race is becoming more consciously aware of itself, its worldwide interconnectedness, and its social and political systems.

Just as when an individual's conscious awareness experiences the ego as a function, collapsing its illusory power, society as a whole is capable of realizing the hollow, empty nature of aggressive, dominating political power structures and becomes incapable of bearing them any longer. When the Egyptian police made the ludicrous claim that Khaled Said had accidentally choked to death, they were playing the same old game they had been playing for generations. But technology and social media had changed the rules. Everyone could see for themselves Said's mutilated face—the photograph had gone viral—and the lie galled. It is one thing to know that the system is rigged; it's quite another for individuals to see the fabrication clearly and to realize that they are being played for fools. In the light of conscious awareness, such duplicity, like the illusion of the ego, cannot survive.

***

Individual development is at the heart of humanity's evolution. How interdependent individual and group evolution are becomes clear when we apply the phase model to human evolution:

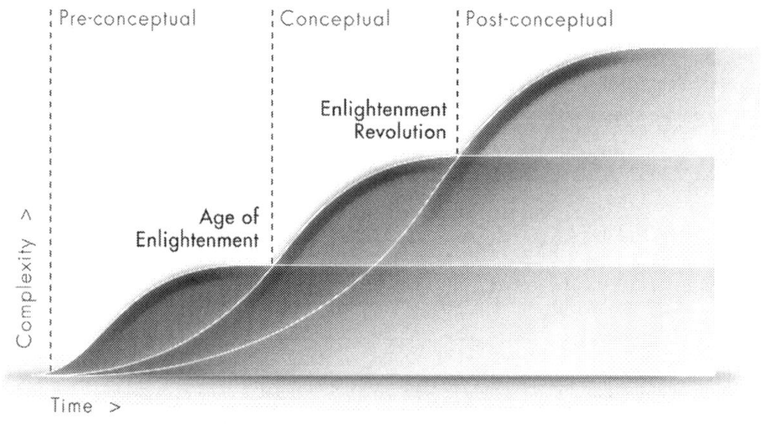

The Phase Model of Human Evolution

The *pre-conceptual phase* of human evolution is akin to the pre-personal phase of development. It was the orientation of our prehuman ancestors. Physical survival was of immediate concern. Violent confrontations, attacks, and defense were an expression of this orientation. A direct, unmediated relationship to sensory awareness was the focus of life. Pain hurt. Food tasted good. The sun was warm. The night was dark. Life unfolded in the immediacy of the moment, with very little notion of a past or a future. The light of conscious awareness was very dim; life was lived mostly subconsciously.

The *conceptual phase* of human evolution is akin to the personal phase of development. It began with the conceptual revolution, fifty thousand years ago. At first these early humans probably had short episodes of symbolic thought, such as when planning a hunt or deciding which field would be best to gather roots in, but these were likely few and far between.

Although this new capacity represented an earth-shattering leap forward, it took millennia for it to gain enough momentum to become our chief mode of orientation. In fact, conceptual orientation probably did not completely outpace physical orientation in the general populace until the Age of Enlightenment in Europe during the eighteenth century. When science usurped religion as the main method of understanding reality, the bulk of humanity for the first time came to experience the personal, conceptual orientation as their main way of relating to life.

During the conceptual phase, cultures exhibit strong national identification. National anthems, flags, languages, cultural heroes, and history are emphasized and celebrated. National boundaries and national interests are protected, often with the threat of force, because we defend self-identity concepts as if our lives depend on them. Ideologies, religious dogmas, and cultural identities become things people will live and die for, as the story of the al-Qaeda terrorists illustrates. Alliances between groups of nations for territorial domination or commerce are an expression of conceptual orientation of larger groups. NATO and the former Soviet bloc are examples.

The *post-conceptual phase* of human evolution is analogous to the post-personal phase of individual development. For most of humanity in the developed world, the wave of the personal orientation will peak within a few generations and be supplanted by the post-personal orientation. This will be reflected in the cultural shift from a conceptual to post-conceptual orientation. Again, this will not be a smooth, linear process but a dynamic one characterized by many expansions and contractions. We are already seeing it in the open source and openhanded culture of the Web, where the goal is to make access available to all, to give away as much as possible for free, to help promote others' work across organizational and national lines.

Conscious orientation probably appeared several thousand years ago in just a few remarkable individuals who were far ahead of the curve. It has slowly but steadily gained traction, and today increasing numbers of people exhibit a greater degree of conscious awareness than ever before. Acceptance of individual human rights—not even on our collective radar a hundred years ago—and respect for cultural differences are expressions of this orientation. The United Nations, the Red Cross, and the International Court of Justice in the Hague are examples of the emerging conscious orientation because they reach beyond the borders of our ego identifications—with countries, races, cultures, religions—and attempt to serve *all* people. Such organizations, even with all their faults and drawbacks, represent the beginning of the social expression of the conscious orientation.

Worldwide, many cultures are in the early transition from conceptual to post-conceptual orientation. Other cultures are still pre-conceptually oriented and transitioning to conceptual orientation. Within each culture there's a broad range of development, but it is the collective state of conscious awareness, the average across the population, that signifies the center of cultural evolution. This is reflected in laws as well as in the degree of openness and tolerance toward differing opinions and perspectives. The more open and accepting a culture is, the closer it moves toward a post-conceptual orientation.

***

The technological and social complexity of the human species is increasing at an exponential rate, and our brain complexity is following suit. The open-mindedness, flexibility, and centeredness that conscious awareness requires are laying the groundwork for it to become the dominant orientation in both individuals and humanity as a whole. When this occurs the human race will enter the next phase of its evolution into a more mature species: a post-conceptual humanity free of the prison of thoughts and feelings, free of the ego's imposed solitary confinement.

When individual awareness expands from identification with a personal self to the post-personal perspective of conscious awareness, the culture as a whole will move from the conceptual to the post-conceptual phase of human evolution. As a species we will be increasingly aware of what is conducive for humans to flourish; we will be willing to take the steps to make it happen for all human beings. This is when the enlightenment revolution will become fully apparent.

The enlightenment revolution will have an impact on humanity as profound as the conceptual revolution has had over the last fifty thousand years. Just as the power of the conceptual revolution lay in its ability to orient us away from the body and toward the conceptual mind, the enlightenment revolution may usher in an era of reintegration of body and mind.

The enlightenment revolution is a silent revolution. It will not be marked by the violent overthrow or even replacement of the older orientations. Instead, these orientations will be integrated into conscious awareness. Both the physical and conceptual capacities are required for an expanded awareness to evolve; thus they are not going to disappear or become unavailable to us.

The early signs of a post-conceptual phase of human evolution are visible. Already the mind-set of the new generation is becoming far more egalitarian. Today's young generation questions authority and dogma, its sexual attitudes are more natural, and its basic mind-set is one of openness and cooperation. Today's political and business leaders exhibit more humanity and less remoteness, such as Steve Jobs's famous stream of personal responses to e-mails and the humorous Twitter feed of Dmitry Medvedev, the president of Russia.

Conscious capitalism is a further indicator of changing times. It is an approach marked by the considerations of all stakeholders: customers, employees, investors, management, and suppliers as well as the environment. Patagonia, Zappos, Whole Foods Market, The Container Store, and countless others approach business as a system where every part needs to be taken into account.

We now live in a world so complex and intertwined that no single brain can know and understand what is actually going on. The last person who had a good working knowledge of all the fields of science was probably Thomas Young, an English scientist and physician who died in 1829. Working together in large groups is the only possible solution, and the coming paradigm will be marked by shared leadership, group activity, and social responsibility. This is not to be mistaken for the reanimated ghosts of socialism or communism. These too were structured as dominance hierarchies and are archaic at their root. The Egyptian revolution is an example of organic cooperation with no apparent leader.

Transparency is another attribute of the emerging consciously oriented culture. In the past, knowledge was power and power meant domination. Secrecy is a fear-based attitude that breeds and perpetuates fear. Today, technology is one of the main drivers toward greater transparency and interconnectedness. WikiLeaks and OpenLeaks are organizations committed to transparency, offering a model that may fundamentally change how government functions. More transparency in a culture leads to more openness. Being consciously aware of what's going on in the world is the best way to a healthier society. As the saying goes, sunlight makes the best disinfectant.

The World Wide Web is probably the most visible expression of the coming changes in social organization. Its key features are egalitarianism, interconnectedness, and openness. These qualities represent a return to our roots, the same key features of social organization that existed during the pre-conceptual phase of our evolution. It is, however, not a regression but a progression because it also includes all the wisdom and benefits accrued during the conceptual phase.

When humanity is steeped in conscious awareness, identification with beliefs and perspectives will vanish from most of the adult population. In the not-too-distant future, for example, our current political discourse will be seen as bad theater, reminiscent of the physical bashing of cavemen. From a post-conceptual perspective, dirty partisan politics and deep nationalism seem like ineffective, old, and tired ways to govern. The fight for physical survival is over, and the fight for intellectual supremacy is on its way out. When ideologies, concepts, and political positions are held lightly, as mere organizing methods rather than personal identities, it will be much easier to find cooperation and work together toward real solutions.

Another expression of shifting times is the current flood of international trade. Historically trade has always been on the forefront of cultural change, and we currently live in the most energized trading environment ever. In the dynamic exchange of international corporations and businesses, national boundaries are beginning to become meaningless. Nationalism is fading, and globalism is on the march. The European Union, although still fragile, is a step beyond national identity. International travel, mixed-race marriages, and multicultural institutions of higher education are paving the way for old rigid identities to melt away.

The spectrum of cultural development globally, however, remains rather broad. Some cultures are still in the transition from a pre-conceptual to a conceptual orientation while others are in the beginning stages of the transition between conceptual and post-conceptual orientation. We do not mean to deny that there are still starvation and oppression in the world and that we still have a long way to go. But today's physical conflicts worldwide are an expression of a pre-conceptual orientation. Stoning a young couple in public as punishment for adultery is sad and unconscious, a reaction mired in the biases and reactivity of the pre-personal orientation. And no matter which way you look at it, even counting mass murder, the proportion of the population dying of violence is nowhere near the one in four who died at the hands of another at the beginning of the conceptual revolution.

Intellectual battles, verbal fights, and smear campaigns in the political arena are symptomatic of the ego's identification with concepts that is at the core of the personal orientation. Yet collectively we're realizing that these conflicts are silly and unproductive. All too often they don't accomplish anything for the whole community and instead serve only the egos of the politicians. As increasing numbers of individuals make the transition from a conceptual to a post-conceptual orientation, this wasteful form of politics will begin to fade away and the process will be oriented more toward results than personalities.

Viewing this as a naïve or overly idealistic proposition is the conceptual mind's perspective, the ego's trap, denying change. Life today would look like heaven from the perspective of a few hundred years ago: from disease prevalence to cleanliness, from ubiquitous violence to physical comfort. Within several generations conscious orientation may very well be as common as conceptual orientation is today.

***

The evolution of our species is happening now, right in front of our eyes. Humanity as a whole is beginning to discover Einstein's "optical delusion of consciousness" and to find relief from our mental prison. The understanding that we are not the mental self-concept we believe ourselves to be is slowly filtering into human awareness. Yet if our ego is not really who we are, a gaping question remains at the center of our subjective experience: Who are we?

Einstein's answer to this question was clear: "A human being is a part of a whole, called by us 'universe,' a part limited in time and space." The significant part here is "...limited in time and space." He didn't propose an immaterial soul or an essence that is immortal. He was referring to a human being as a life form that is born and eventually dies. The unvarnished truth is that each of us is an individual biological organism subject to the biology that governs our capacity to stay alive.

The human brain evolved the capacity to be consciously aware. We know that we exist and that we will face death at some point. The certainty that we will one day die has been humanity's biggest mental hurdle. No other biological organism on earth, not even our closest relatives, the great apes, possesses the ability to imagine its future death. They have never had to face this reality, and evolution has not had sufficient time to prepare conceptually aware biological organisms to deal with it.

In attempting to come to terms with the distressing certainty of our mortality, we invented concepts such as heaven, the afterlife, reincarnation, and angels. The idea of an immaterial soul that would survive bodily death was a creative way to deal with the frightening prospect of dying. Such belief systems have proven to be adaptive and were probably the best thing natural selection could come up with at the time.

Yet it seems evolution may be forming a new solution. Becoming consciously aware tears down the walls of the mental prison that isolates us from feeling fully alive. It allows us to recognize ourselves as human beings limited in time and space but with the capacity to be consciously aware of being alive. The full knowledge that we're going to die doesn't take away from that inherently rewarding experience available in every moment: it makes it all the more valuable. To the post-personal orientation, the concept of a future death is no longer personally terrifying. It becomes just another idea; true, but no cause for worry. Yes, the organism will die, but the ego that is terrified of this is illusory.

It is humbling to consider our vulnerability as living organisms. The ego does nothing to keep the body functioning properly; it all happens subconsciously. Our lungs know how to breathe and take in oxygen while expelling carbon dioxide; our intestines know how to digest food. The notion that there's an egoic entity in control of our lives is a concept that stems from identification with a mental self-concept. Being consciously aware of the constructed nature of the ego leads to the inevitable conclusion that there's no personal achievement involved.

Our lives are limited in time and space, yet they provide the fertile ground for following generations. To be consciously aware that we're each a piece of the puzzle of humanity's evolution can be humbling but it can also be a big relief. The illusion of being a separate self creates an enormous burden on us. The realization that life is not about us personally, that life is a process beyond our individual existence, frees us up to participate more fluidly and more fully. Liberation from identification, from toxic egotism, allows us to feel at home in the universe, part of the ongoing process of the evolution of our species.

***

Organisms adapt to their environment. But thanks to our conceptual capacity we're changing our environment as much as our environment is changing us. We're collectively becoming more consciously aware as our culture evolves. We've become aware of the dangers of smoking tobacco, so we try to prevent it and encourage each other to refrain. We instituted speed limits because we became conscious of the dangers of driving too fast. Our manipulation of the environment is geared toward expanding physical safety and comfort. However, we're reaching a point of diminishing returns in the physical realm. Increasing conscious awareness will allow us to improve our inner mental and emotional environment.

Up to this point, it was a conceptual norm to blame and punish, to praise and reward. This established a social norm of behavior that was conducive to the expansion of our civilization. Individuals became a resource to be maximized to ensure the survival of all. No one designed this mechanism; it evolved. The extreme case of using human beings as resources was slavery. In blind and ignorant fashion, the focus was on how to get the job done, regardless of the impact on the individual. This is changing dramatically. In today's world the well-being of all our citizens is a major concern.

As our evolution continues, we may someday perceive each other not as separate and potentially dangerous individuals but as fellow travelers, each equally vital to this mind-boggling process we call life. Just as we have discovered that our planet is not the center of the universe, soon we may realize that our experience as a separate ego is an equally mistaken perspective. In modern society most of us now enjoy freedom of association, freedom of expression, freedom of movement, and freedom of speech. For most of us, however, freedom from identification is still unknown. The boundaries created in the conceptual mind will be the next barrier that evolution breaks down.

As human beings evolved, we learned about our bodies, gravity, the genetic code, quantum uncertainty, and black holes. Eventually we will fully understand the nature of our minds and of illusions that they generate. Once we completely comprehend this illusion, the experience of life will never be the same. The time may come when the mental prison of the ego-construct will be as much a relic of our evolutionary past as the stone chopper is of the end of our pre-personal phase.

Life as a whole is so intertwined and interactive that nothing is separate and no one independent. No individual is expendable, yet no one is vital or in control of this interconnected web of life. Evolution often seems brutal and unkind. The survival of the fittest feels like a distressing and inhumane existence. The fear, pain, and sorrow inherent in this struggle are known to us all. The larger context, however, can also be seen as magnificent. Life is an inexplicable, unstoppable force, resilient enough to overcome the depth of despair and violent creation, continuously bootstrapping itself toward higher levels of manifestation.

Liberation from the illusion of being a separate self is a natural event, no more or less special than a flower opening toward the sun or an infant recognizing its mother's face. We will realize Einstein's claim that our individual lives are an essential part of a larger process, a universe in motion, and in that realization we will find freedom from the human condition.

peter baumann's acknowledgments

Many acknowledgements are due to all of my friends who indulged me over the years, listening to my latest musings. Thank you for your patience and your feedback.

Michael, you know this book would not have made it without your relentless research and your incredible efforts in shaping a diverse set of ideas into a coherent sequence, and finally putting it all into writing; my deepest gratitude to you.

Thank you, Penelope, for your support in this process.

Amy Hertz, it's still shocking to me how you were able to fly in from New York and so boldly and decisively help us edit the material. Your sharp focus and bold style have made a dramatic difference. Thank you.

Jeff, thank you for your steady encouragement and never-ending support for more than two decades now. Can you believe it? The best is yet to come!

Tami, thank you for being such an incredible person; I enjoy our journey together.

I have to extend a special thanks to you, Gil Evans, who for years have been a steady sounding board and collaborator.

Willa, Ahmet, Max, and Ema, you are the next generation. I'm so proud of who you are; I love you dearly.

Alison, you've always overlooked my shortcomings and never wavered in your encouragement and support. No words can describe the depth of love I feel for you.

michael w. taft's acknowledgments

To Robert Nash, in memoriam.

Thanks to all the people who did so much behind the scenes to make this book a reality, including Alison Baumann, Jeff Klein, Catherine Hollis, Jenna Young, and Lindsay Starke.

Thank you Amy Hertz: your wit, wizardry, and razor-sharp red pen transmogrified a stack of words into a book. Idea machines are fun!

Thank you Morgan Blackledge, for consistently and brilliantly charting the unknown with me.

Thank you Tami Simon, for your fierce friendship, impeccable matchmaking abilities, and supreme authenticity.

Peter: thank you for the opportunity to collaborate with you on this book and for fostering an open and intelligent atmosphere to exchange ideas. Thank you most of all for a great friendship. It's all just happening.

Thanks to Dhyanyogi Madhusudandasji, Anandi Ma Pathak, and Shinzen Young for their inspiration and guidance over the years. *Atta dipa viharatha.*

And most of all, thank you Penelope for being the love of my life.

notes

These notes will be continuously expanded and updated on the website www.egothebook.com.

Chapter 1: Evolution's Unfinished Product

"...nineteen hundred family homes.": Taylor, 2010. According to the National Association of Home Builders, the floor area of the average American home is 2,309 square feet.

"...water treatment plant.": Boeing, 2011.

"...three hundred thousand houses for a month.": US Energy Information Administration, 2011. One billion BTU equal all the electricity that three hundred households consume in one month.

"...fifty miles in clear weather.": Darton, 1999.

"...digital television mast.": Ibid.

"...cobblers of the colony.": New York City Fire Department, 2011.

"...rushed to the area.": McKinsey Report, 2002.

"...fifteen thousand images.": The September 11 Digital Archive, 2004.

"...'Down with America'.": Fox News, 2001.

"...Turkish, and other militaries.": Operation Enduring Freedom, 2011.

"...their lives in Afghanistan.": Highwater, 2010.

"...at the hand of another.": Wade, 2006.

"...to kill or to die.": Ibid.

"...reflect each other's state of mind.": Gallese, 2001.

"...killed without sin.": Qutb, 2003.

"...happening in the cabin.": Moyers, 2007.
"...would suddenly go dry.": Fink, 2000, p.408.
"...whole of nature in its beauty.": Einstein, 1950.

Chapter 2: What Are Emotions For?

"...I am home and safe.'": Wajda, 2001.
"...Fireworks really scare me now.'": Belluck, 2002.
"...that his tongue was his tongue": Gajilan, 2006.
"...would have been very short.": Ibid.
"...he's almost eight years old.": "Help Roberto," 2007.
"...paralyzed by anxiety and lethargy.": Portenoy et al, 1986.
"...a comfortable homeostasis.": Damasio, 2003.
"...we are emotional lizards.": LeDoux, 1998, p.174.
"...learnt by the individual.'": Darwin, 1898.
"...puny desert bugs they are.'": Axelrod, 2001.
"...occur IS the emotion.'": James, 1890, p.449.
"...in the body that shapes emotion.": Sapolsky, 2010.
"...drug's effect on their body.": Schacter and Singer, 1962.
"...culture attempted to conceal.": Ekman and Friesen, 1975.
"...or vomit something out.": Chapman, 2009.

Chapter 3: The Emotional Nervous System

"...employees in the office was alive.": Gordon, 2001.
"...another one of them.'": Homans, 2003.
"...Welcome to my world.'": Finn, 2002.
"...a kind of opiate-withdrawal state.": Panksepp, 1998.

"...hypothalamus to control body functions.": Sapolsky, 2010.
"...dopamine, norepinephrine, and phenylethylamine": Crooks and Baur, 2008, p.171.
"...creating a dependency-like bond": Ibid., p.172.
"...preference for certain types of sweat.": Wyart et al, 2007.
"...is the most different from their own.": Lie et al, 2010.
"...strong, healthy children.": Berreby, 1998.

Chapter 4: The Pursuit of Happiness

"...we did not care.'": Mazzocchi, n.d.
"...nothing could bother him.": Kreider, 2009.
"...make us feel really good.": Lykken, 1996.
"...by evolution to live.": Grinde, 2002.
"...frustrations again possessed him.": Kreider, 2009.
"...than a million.'": Daily Mail, 2011.
"...or hedonic tone.'": Brickman and Campbell, 1971.
"...accident on people's happiness.": Brickman, 1978.
"...decisions of happy people.": Lerner and Keltner, 2001.
"...tempting humans in these sins.": Binsfield, 1589.

Chapter 5: Why Do I Care?

"...treasure the rest of my life.": O'Keefe, 2004.
"...believe in me again.": Fox News, 2010.
"...power of embarrassment.": Semin and Manstead, 1982.
"...emotions of other people.": Iacoboni, 2008.
"...inaccessible to experiments.": Ramachandran, n.d.

"...and inform their behavior.": Adolphs, 2002.
"...hasn't ended for any of us.'": Domash, 2002.
"...for any spelling errors.'": Olszewski, 2010.
"...who were born blind revealed this.": Tracy and Matsumoto, 2008.

Chapter 6: Becoming Human

"...dreamed up the September 11 attacks.": CNN, 2002.
"...forced to flee.": McDermott, 2002.
"...all the women and children.": Wright, 2006.
"...training camp near Kandahar.": 9-11 Commission, 2004. See chapter 5.
"...relinquish world power.": Bin Laden, 2003.
"...get stranded and die.": National Geographic, 2007.
"...back into the water.": Heimlich, 2001.
"...delicate manipulation.": Foley, 2004.
"...their families now.": Frederick, 2001.
"...at the operational level.": Dwyer, 2002.
"...is just fine with me.": Frederick, 2001
"...departments and agencies.": US Department of Labor, 2010. This does not include postal workers, for some reason.
"...level of a toddler.": Walker and Shipman, 1996.
"...to survive as long as he did": Rasmussen, 1993.
"...to live several more years." Graham, 2005.
"...to read in the bones": Banathy, 2000.

Chapter 7: The Conceptual Revolution

"...'our society,' said Jones.": Jones, 2010.
"...world domination.": Robertson, 2009.

"...Mohammed to Hitler.": Robertson, 2002.
"...a religion of war.": Graham, 2010.
"...all living human beings.": Islamic Population, 2009.
"...dig anywhere in the world.": Thomas, 2006.
"...Europe at the time.": Richards, 2000; Leonard, 2002.
"...freshwater fish regularly.": Hong, 2007.
"...more efficient neurons.": Deans, 2011.
"...the first fishermen.": Crawford, 1999. Although it may not have been necessary, see Carlson, 2007.
"...smaller, more manageable groups.": Dunbar, 1997.
"...capable of modern language.": Wade, 2006. The FOXP2 gene.
"...Sally doesn't know what they know": Wimmer and Perner, 1983.
"...different from their own": Premack and Woodruff, 1978.
"...thus religion was born.": Atran, 2002.
"...in their most potent form.": Boyer, 2001.

Chapter 8: Prima Donna

"...the Saudi Binladen company.": Wright, 2006. The name bin Laden is spelled many different ways in English.
"...fathered around fifty-five children": bin Laden, 2009. The number of children listed varies. It seems to be somewhere in the mid-fifties.
"...worldwide fundamentalist Islamic theocracy.": Springer, 2009.
"...was with him when he died.": bin Laden, 2009; Cahalan, 2009; Wright, 2006.

"...meaning of individuality and identity.": With the notable exception of the Buddha.
"...duplicate structures in the brain.": LeDoux, 2002.
"...all major researchers in neurology.": Including LeDoux, Gazzaniga, Damasio, Dennett, Wegner, and many others.
"...self-concept as a network of associations.": Gazzaniga, 2009.
"...body awareness and emotional tone.": Damasio, 2003.
"...synapses of the brain, virtually everywhere": LeDoux, 2002.
"...self-referential thoughts and feelings.": Leary and Buttermore, 2003; Metzinger, 2009.
"...rudiments of a mental self-concept.": Cartwright, 2002.
"...they made him god and one third man.": AINA, 2011
"...a city-state in modern Iraq.": From which we get the name "Iraq."
"...an estimated thousand tons.": Mithen, 2004.
"...Tenno, or 'Heavenly Sovereign.'": Stockwin, 1999.
"...making up the empire of Egypt.": Kerr, 2008. Population statistics on ancient Egypt are approximate.
"...but they did not explode.'": Woolf, 2004.
"...to make His religion victorious.'": Bergen, 2006.
"...a litter of puppies his boys had adopted.": bin Laden, 2009.
"...pick letters from their own name.": Nuttin, 1985.

"...pronouns, 'they,' 'them,' 'theirs.'": Perdue et al, 1990, Otten and Wentura, 1999.
"...something called the "endowment effect.": Thaler, 1980. Term coined by researcher Richard Thaler.
"...watching a college basketball game.": University of Utah, 1998.

Chapter 9: Wake-Up Call

"...this 115th shuttle mission": NASA, 2006. Although it was the 115th mission, its designation was STS-121.
"...because of her sunny personality.": Schneider, 2006.
"...It went off without a hitch.": NASA, 2006.
"...killing and eating each other": Wade, 2006. As derived from both genetic and fossil evidence.
"...the explosion in toolmaking.": Ibid.
"...religious texts, and scholarly works.": Schmandt-Besserat, 2002. Writing was also invented independently in China, where it evolved from symbols scratched into turtle shells used in divination.
"...Elijah, Isaiah, and Zoroaster.": Beversluis, 2000.
"...of two million under King Menes.": Shaw, 1995, p.560. Menes is a Greek name, probably referring to the Pharoah Narmer.
"...64 million individuals.": Hansen, 2006. The population of the Greek empire in the fourth century BCE was from 8 million to 10 million people, as estimated by Mogens Hansen. The population of the ancient Egyptian empire at its highest, during the New Kingdom, was 2.1 million around 1300BCE.

(Later populations were greater, but this was during Roman times.) See Shaw, 1995.

"...population of about a million.": Hopkins, 2008.

"...a staggering 1.6 billion.": Rosenberg, 2011.

"...one of whom would die in its first year.": Centers for Disease Control, 1999.

"...nearly one in a hundred": Ibid. The figure given is six to nine out of 1,000.

"...forgotten for another fifty years.": Magner, 2005, p.311. Semmelweis had this insight in 1847, but doctors resisted the idea because there was no germ theory. When this was developed (mainly by Pasteur), it was initially resisted, but "...by about 1895 this opposition [to germ theory] was essentially disarmed by the dazzling prospects of powerful new therapeutic tools that actually strengthened the medical profession."

"...subspecialities for doctors alone.": American Board of Medical Specialties, 2010.

"...procedures number around eight thousand.": The AMA's list of CPTs (current procedural terminology).

"...twenty-four thousand prescription medications.": Drugs. com, 2011.

"...expectancy has doubled since 1850.": Infoplease.com, 2011. From 38.3 to 75.7 in men, and from 40.5 to 80.8 in women.

"...time to be funding spacecraft.": Martino, 1992, p.43.

"...on returning from the moon.": Cunningham, 2002. Thomas Stafford, John W. Young, and Eugene Cernan set the record for the highest speed attained by a manned vehicle at 39,897 km/h (11.08 km/s or 24,791 mph) during the return from the Moon on May 26, 1969.

"...bridging cultures and widening trade.": Ridley, 2010.

"...carrying around 56 million people.": RITA, 2003.

"...Kurzweil has mapped this rate of change.": Kurzweil, 2001.

"...apprehension and interpretation of reality.": Pogge, 2009.

"...live television right on the street.": Jennings, 2001.

"...made worse by several fires.": The 9-11 Commission, 2004.

"...in history to collapse from fire.": Griffin, 2010. This is only one of numerous conspiracy books to repeat this meme.

"...means of promoting an agenda.": Avery, 2005–2009.

"...Jennings mysteriously died.": Dykes, 2008.

"...which contradicted his statements.": NIST, 2008.

"...as did viewers from Dartmouth.": Hastorf, 1954.

"...not react fast enough to danger.": Fiske, 1991.

"...but you see *with* it.'": Metzinger, 2009.

"...covered in zippers and snaps.": Fox News, 2008.

"...Huthaifa Azzam": Wright, 2006. Son of cleric Abdullah Azzam, the man most responsible for turning bin Laden toward violence.

"...What did we get from September 11?'": Bergen, 2007.

"...where only civilians are dying.": CNN, 2008.
"...Indonesian branch of al-Qaeda": Named Jemaah Islamiya.
"...began working with the police.": Mydans, 2008.
"...publicly repudiate al-Qaeda.": Bergen, 2008.
"...commander of al-Qaeda in Algeria": Known as "al-Qaeda in the Islamic Maghreb."
"...he was waging was illegitimate.": Memri, 2008.
"...rejecting the group's methods.": Wright, 2008.
"...recordings of the 7 WTC collapse.'": NIST, 2008, 38–39.
"...I honestly don't believe that.'": BBC, 2008.
"...from the site just in time.": NIST, 2008.
"...a storm broke loose'": Rigden, 2006.
"...towards a theory of gravitation.'": Ibid., p.5.

Chapter 10: The Organism Is in Charge

"...like the major companies do.": "History Commons," 2001.
"...helping to kill thousands of people.": Goldstein and Finn, 2001.
"...if someone hurt an insect.": Cloud, 2001.
"...LeDoux calls the 'emotional lizard.'": LeDoux, 2002.
"...consciously thinking about how they did it.": Hélie and Sun, 2010
"...all these plans without telling you about it.": Lehrer, 2008.
"...as science writer Nicholas Wade remarks.": Wade, 2006.

"...it's happening too slowly to notice.": ABC-TV, 2004.

"...average IQ back then would only be 80!": Neisser, 1997.

"...generation of children was doing better on IQ tests.": Flynn, 1984.

"...one example is Siddhartha Gautama.": The historical person known as the Buddha, meaning "Awakened One."

"...a mirage, a shadow, and dream.'": Lin, 1955.

Chapter 11: Developing Enlightenment

"...had this experience at twenty-eight.": James, 1902. Although he attributes it to a letter from an "anonymous Frenchman," posterity has concluded that James was actually describing his own experience.

"...the mechanistic theories of the German psychologists.": Johann Herbart, for example.

"...epileptic patient he had seen paralyzed him.": Hunt, 1993.

"...groundbreaking classic, The Principles of Psychology.": James, 1890.

"...noticeable through conscious awareness.": He labeled this "pure Ego," a slightly confusing term for our purposes.

"...recedes into the tailbone.": Garstang, 1922.

"...move on to Shakespeare and algebra.": Davidson, 1914; Fox, 1925.

"...scientific background and understanding for them.": For example, Piaget, Lovinger, and Kohlberg.

"...three or four years before we have 20/20 vision.": Shelov, 2009.

"...and quadruple by twenty-four months.": "Infancy," 2011.

"...'Mommy' and 'Daddy' fairly quickly.": Ibid.

"...from misery to 'ordinary unhappiness.'": Freud and Breuer, 1895.

"...live in a more peaceful existence.": Wahab, 2006.

Chapter 12: The Rise of an Enlightened Humanity

"...'We are all ready to die.'": Ghonim, 2011.

"...and reinforced by many others.": Kirkpatrick, 2011.

"...'no one is the hero in that picture.'": Scola, 2011.

"...oppression broke down completely.": Cohen, 2011.

"...physician who died in 1829.": Although some claim it was Athanasius Kircher, who died in 1680

"...the forefront of cultural change.": Ridley, 2010.

references

The 9-11 Commission. *The 9-11 Commission Report: Final Report of the National Commission on Terrorist Attacks upon the United States.* New York: Norton, 2004.

ABC-TV. "Catalyst: IQ," March 18, 2004. Retrieved from http://www.abc.net.au/catalyst/stories/s1068478.htm.

AINA. "The Epic of Gilgamesh," 2011. Retrieved from http://www.aina.org/books/eog/eog.htm.

"Ancient Egypt." "Geography of Ancient Egypt," 2011. Retrieved from http://www.experience-ancient-egypt.com/geography-of-ancient-egypt.html.

Adolphs, R. et al. "Impaired Recognition of Social Emotions Following Amygdala Damage." *Journal of Cognitive Neuroscience,* 14 (8), 1264–74, November 15, 2002.

American Board of Medical Specialties. "Specialities and Subspecialities," 2010. Retrieved from http://www.abms.org/who_we_help/physicians/specialties.aspx.

Atran, Scott. *In Gods We Trust: The Evolutionary Landscape of Religion.* Oxford: Oxford University Press, 2002.

Avery, Dylan. "Loose Change 9-11: An American Coup, 2005–2009." Retrieved from http://www.loosechange911.com/.

Axelrod, Steven. "9/11 letters: Rage, Hope and the American Jihad." Open Salon, September 9, 2010. Retrieved from http://www.open.salon.com/blog/steven_axelrod/2010/09/09/911_letters_rage_hope_and_the_american_jihad.

Banathy, B.H. *Guided Evolution of Society: A Systems View.* New York: Kluwer Academic/Plenum Publishers, 2000.

BBC News. "The Third Tower: Barry Jennings," 2008. Retrieved from http://www.youtube.com/watch?v=hsOKkDwyfjg.

Belluck, Pam. "With Patriotism Renewed, July 4 Hits a Deeper Chord." *The New York Times,* July 4, 2002.

Bergen, Peter L. *The Osama Bin Laden I Know: An Oral History of al Qaeda's Leader.* New York: Free Press, 2006.

Bergen, Peter. "War of Error." *The New Republic,* October 22, 2007.

Bergen, Peter. "The Unraveling." *The New Republic,* June 11, 2008.

Berreby, David. "Studies Explore Love and the Sweaty T-shirt." *The New York Times,* June 9, 1998.

Beversluis, Joel, ed. *Sourcebook of the World's Religions,* 3rd ed. Novato, CA: New World Library, 2000.

bin Laden, Omar and Najwa bin Laden. *Growing Up bin Laden: Osama's Wife and Son Take Us Inside*

Their Secret World. New York: St. Martin's Press, 2009.

bin Laden, Osama. "Osama Bin Laden's Sermon for the Feast of the Sacrifice." *The Middle East Media Research Institute,* No. 476, March 6, 2003. Retrieved from http://www.memrijttm.org/content/en/report.htm?report=822&param=JT.

Binsfield, Peter. "Correspondence of Demons to the Seven Deadly Sins," 1589. Retrieved from http://thegrimoire.tumblr.com/post/1034630369/correspondence-of-demons-to-the-seven-deadly-sins.

Boeing. "History of Boeing and the Everett site," 2011. Retrieved from http://www.boeing.com/commercial/facilities/index.html.

Boyer, Pascal. *Religion Explained: The Evolutionary Origins of Religious Thought.* New York: Basic Books, 2001.

Brickman, Philip and Donald Campbell, M.H. Apley, ed. "Hedonic Relativism and Planning the Good Society." *Adaption Level Theory: A Symposium.* New York: Academic Press, 1971.

Brickman, Philip et al. "Lottery Winners and Accident Victims: Is Happiness Relative?" *Journal of Personality and Social Psychology,* 36 (8), 917–27, August 1978.

Cahalan, Susannah. "Tales from the bin Laden Clan." *The New York Post,* October 11, 2009.

Carlson, B.A. et al. "Docosahexaenoic Acid Biosynthesis and Dietary Contingency: Encephalization without Aquatic Constraint." *American Journal of Human Biology,* 19 (4), 585–8, July–August, 2007.

Cartwright, Jo-Anne. *Determinants of Animal Behaviour.* London: Routledge, 2002.

Centers for Disease Control. "Achievements in Public Health, 1900–1999: Healthier Mothers and Babies," 1999. Retrieved from http://www.cdc.gov/mmwr/preview/mmwrhtml/mm4838a2.htm.

Chapman, Hanah et al. "In Bad Taste: Evidence for the Oral Origins of Moral Disgust." *Science* (323: 5918), 1222–1226, February 27, 2009. Retrieved from http://www.sciencemag.org/content/323/5918/1222.full.

Cloud, John. "Atta's Odyssey." *Time,* September 30, 2001. Retrieved from http://www.time.com/time/magazine/article/0,9171,176917-1,00.html.

CNN. "September 11 Mastermind Graduated from U.S. University." CNN.com/US, December 19, 2002. Retrieved from http://edition.cnn.com/2002/US/South/12/19/al.qaeda.aggie/.

CNN. "Bin Laden's Son to Father: Change Your Ways," January 21, 2008. Retrieved from http://articles.cnn.com/2008-01-21/world/binladen.son_1_bin-laden-father-osama-bin?_s=PM:WORLD.

Cohen, Noam. "Egyptians Were Unplugged, and Uncowed." *The New York Times,* February 20, 2011.

Crawford, Michael. "Evidence for the Unique Function of Docosahexaenoic Acid during the Evolution of the Modern Hominid Brain." *Lipids* 34, 1999.

Crooks, Robert and Karla Baur. *Our Sexuality,* 10th ed. Belmont, CA: Thomson Wadsworth, 2008.

Cunningham, Antonia, ed. *Guinness World Records 2002.* New York: Bantam Books.

The Daily Mail. "King of Chavs Has to Do a Lotto Work," March 18, 2011. Retrieved from http://www.dailymail.co.uk/news/article-1367178/Michael-Carroll-King-Chavs-Lotto-workpunishment-drink-driving.html.

Damasio, Antonio. *Looking for Spinoza: Joy, Sorrow, and the Feeling Brain.* New York: Harcourt, 2003.

Darton, Eric. *Divided We Stand: A Biography of New York's World Trade Center.* New York: Basic Books, 1999.

Darwin, Charles. *The Expression of Emotions in Man and Animals,* 3rd ed. New York: D. Appleton and Company, 1898.

Davidson, P.E. *Recapitulation Theory and Human Infancy.* New York: Teacher's College, 1914.

Dawkins, Richard. *The Greatest Show on Earth: The Evidence for Evolution.* New York: Free Press, 2009.

Deans, Emily. "Your Brain on Omega 3." *Psychology Today,* March 17, 2011.

Dennett, Daniel. *Consciousness Explained.* New York: Back Bay Books, 1991.

Domash, Shelly Feuer. "9/11: One Year After." *Police,* September 1, 2002. Retrieved from http://www.policemag.com/Channel/SWAT/Articles/2002/09/One-Year-After/Page/2.aspx.

Drugs.com, 2011. Retrieved from http://www.drugs.com.
Dunbar, Robin. *Grooming, Gossip, and the Evolution of Language.* Cambridge: Harvard University Press, 1997.
Dwyer, Jim et al. "9/11 Exposed Deadly Flaws in Rescue Plan." *The New York Times,* July 7, 2002.
Dykes, Aaron. "Key Witness to WTC Explosions Dead at 53," 2008. Retrieved from http://www.infowars.com/key-witness-to-wtc-7-explosions-on-911-dead-at-53/.

Einstein, Albert. "Letter to Mr. George Salit." Albert Einstein Archives, Document 61: 226. Jerusalem: Hebrew University of Jerusalem, March 4, 1950.
Ekman, Paul and Wallace V. Friesen. *Unmasking the Face: A Guide to Recognizing Emotions from Facial Expressions.* Englewood Cliffs, NJ: Prentice Hall, 1975.
Fink, George, ed. *Encyclopedia of Stress.* San Diego, CA: Academic Press, 2000.
Finn, Robin. "Public Lives: Working through Pain to Honor a Brother's Life." *The New York Times,* February 26, 2002.
Fiske, S.T. and S. E Taylor. *Social Cognition,* 2nd ed. New York: McGraw-Hill, 1991.
Flynn, J.R. "The Mean IQ of Americans: Massive Gains 1932–1978." *Psychological Bulletin,* 95, 25–51, 1984.

Foley, Robert Andrew and Roger Lewin. *Principles of Human Evolution.* Oxford: Blackwell Science, Ltd, 2004.

Fox, Charles. *Educational Psychology.* London: Keegan Paul, 1925.

Fox News. "Arafat Horrified by Attacks, but Thousands of Palestinians Celebrate; Rest of World Outraged," September 12, 2001. Retrieved from http://www.foxnews.com/story/0,2933,34187,00.html.

Fox News. "Osama bin Laden's Son Says He Wants to Be 'Ambassador for Peace' between Muslims and the West," January 18, 2008. Retrieved from http://www.foxnews.com/story/0,2933,323605,00.html.

Fox News. "Tiger Woods Press Conference," February 19, 2010. Retrieved from http://www.youtube.com/watch?v=pahxZnEyMNI.

Frederick, Sue. BBC News, 2001. Retrieved from http://news.bbc.co.uk/hi/english/static/in_depth/americas/2001/day_of_terror/eyewitness/5.stm.

Freud, Sigmund and Josef Breuer. *Studies in Hysteria.* New York: Penguin Classics, 2004.

Gajilan, A. Chris. "World without Pain Is Hell, Parent Says." *CNN Health,* January 27, 2006. Retrieved from http://articles.cnn.com/2006-01-27/health/rare.conditions_1_roberto-salazar-tongue-autonomic-nervous-system?_s=PM:HEALTH.

Gallese, Vittorio. "The 'Shared Manifold' Hypothesis: From Mirror Neurons to Empathy." *Journal of Consciousness Studies* (8): 33-50, 2001.

Garstang, W. "The Theory of Recapitulation: A Critical Re-statement of the Biogenetic Law." *Journal of the Linnean Society of London,* 35 (232), 81–101, 1922.

Gazzaniga, Michael S. Human: *The Science Behind What Makes Your Brain Unique.* New York: HarperCollins, 2009.

Ghonim, Wael. Twitter feed, 2011. Retrieved from h ttps://twitter.com/#!/Ghonim.

Gjenvick-Gjønvick Archives. "Provisioning a Transatlantic Liner," 2011. Retrieved from http://w ww.gjenvick.com/SteamshipArticles/Provisions/190 1-06-29-ProvisioningATransatlanticLiner.html.

Goldstein, Amy and Peter Finn. "Hijack Suspects' Profile: Polite and Purposeful." *The Washington Post,* September 14, 2001.

Gordon, Meryl. "Howard Lutnick's Second Life." *New York Magazine,* December 11, 2001.

Graham, Franklin. CNN Politics, 2010. Retrieved from http://politicalticker.blogs.cnn.com/2010/08/19/gra ham-obama-born-a-muslim-now-a-christian/.

Graham, Sarah. "Toothless Skull Raises Questions about Compassion among Human Ancestors." *Scientific American,* 2005. Retrieved from http://w ww.scientificamerican.com/article.cfm?id=toothless -skull-raises-qu.

Griffin, David Ray. *The Mysterious Collapse of World Trade Center 7.* Northampton, MA: Olive Branch Press, 2010.

Grinde, Bjørn. *Darwinian Happiness: Evolution as a Guide for Living and Understanding Human Behavior.* Princeton, NJ: The Darwin Press, Inc., 2002.

Hansen, Mogens Herman. *The Shotgun Method: The Demography of the Ancient Greek City-State Culture.* Columbia and London: University of Missouri Press, 2006.

Hastorf, A.H. and H. Cantril. "They Saw a Game: A Case Study." *Journal of Abnormal and Social Psychology,* 49, 129–34, 1954.

Heimlich, Sara and James Boran. *Killer Whales.* Stillwater, MN: Voyageur Press, 2001.

Helen and Harry Highwater. "Unknown News," 2010. Retrieved from http://www.unknownnews.net/casualties.html.

Hélie, S. and R. Sun. "Incubation, Insight, and Creative Problem Solving: A Unified Theory and a Connectionist Model." *Psychological Review,* 117, 994–1024, 2010.

"Help Roberto," 2007. Retrieved from http://www.helproberto.com/index.php.

History Commons. "Profile: Brad Warrick," 2001. Retrieved from http://www.historycommons.org/entity.jsp?entity=brad_warrick_1.

Homans, John. "Tears of a CEO." *New York Magazine,* February 3, 2003.

Hong, Shang et al. "An Early Modern Human from Tianyuan Cave, Zhoukoudian, China." *Proceedings of the National Academy of Sciences,* 104 (16), 6573–78, April 17, 2007.

Hopkins, Keith, Morris and Schiedel, eds. "The Political Economy of the Roman Empire." *The Dynamics of Ancient Empires,* Oxford University Press, 2007.

Hunt, Morton. *The Story of Psychology.* New York: Anchor Books, 1993.

Iacoboni, Marco. *Mirroring People: The New Science of How We Connect with Others.* New York: Farrar, Straus and Giroux, 2008.

"Infancy—Physical Development," 2011. Retrieved from http://social.jrank.org/pages/333/Infancy-Physical-Development.html.

Infoplease.com. "Life Expectancy by Age: 1850–2004," 2011. Retrieved from http://www.infoplease.com/ipa/A0005140.html.

Islamic Population. "Muslim Population Worldwide," 2009. Retrieved from http://www.islamicpopulation.com/index.html.

James, William. *The Principles of Psychology,* Vol. 2. New York: Henry Holt and Company, 1890.

James, William. *The Varieties of Religious Experience.* London: Longmans, 1902.

Jennings, Barry. "9/11 Early Afternoon ABC-7 Interview," 2001. Retrieved from http://www.youtube.com/watch?v=5LO5V2CJpzI.

Jones, Terry. Dove World Outreach Center, 2010. Retrieved from http://www.doveworld.org/the-sign.

Kemp, Barry J. *Ancient Egypt: Anatomy of a Civilization.* London: Routledge, 1989.

Kerr, Adrian. Ancient Egypt and Us: *The Impact of Ancient Egypt on the Modern World.* Fort Myers, FL: Ferniehirst Publishing, LLC, 2008.

Kirkpatrick, David D. "Google Executive Who Was Jailed Said He Was Part of Facebook Campaign in Egypt." *The New York Times,* February 27, 2001.

Kreider, Tim. "Reprieve." *The New York Times,* June 2, 2009.

Kurzweil, Ray. "The Law of Accelerating Returns," 2001. Retrieved from http://www.kurzweilai.net/the-law-of-accelerating-returns.

Leary, M.R. and N.E. Buttermore, N.E. "Evolution of the Human Self: Tracing the Natural History of Self-awareness." *Journal for the Theory of Social Behavior,* 33, 365–404, 2003.

LeDoux, Joseph E. *The Emotional Brain: The Mysterious Underpinnings of Emotional Life.* New York: Touchstone, 1998.

LeDoux, Joseph E. *Synaptic Self: How Our Brains Become Who We Are.* New York: Viking, 2002.

Lehrer, Jonah. "The Eureka Hunt." *The New Yorker,* July 28, 2008.

Leonard, William R. "Food for Thought: Dietary Change Was a Driving Force in Human Evolution." *Scientific American,* November 13, 2002.

Lerner, Jennifer and Dacher Keltner. "Fear, Anger, and Risk." *Journal of Personality and Social Psychology,* 81(1), 146–59, July 2001.

Lie, H.C. et al. "Is Genetic Diversity Associated with Mating Success in Humans?" *Animal Behavior,* 79, 903–09, 2010.

Lin, Yutang, ed. *The Wisdom of China and India.* New York: Modern Library, 1955.

Lykken, David and Auke Tellegen. "Happiness Is a Stochastic Phenomenon." *Psychological Science,* 7 (3), 1996. Retrieved from http://www.psych.umn.edu/psylabs/happness/happy.htm.

Magner, Lois N. *A History of Medicine,* 2nd ed. Boca Raton, FL: Taylor & Francis Group, LLC, 2005.

Martino, Joseph Paul. *Technological Forecasting for Decision Making,* 3rd ed. New York: McGraw-Hill, 1992.

Mazzocchi, Frank. "Sanitation Officers Local 444 SEIU Salutes the Heroes and Mourns the Victims of the Attack on America 9/11/2001." Local 444 SEIU, (n.d.). Retrieved from http://www.local444seiu.com/survivor.asp.

McDermott, Terry. "The Plot." *Los Angeles Times,* September 1, 2002.

McKinsey Report. "Increasing FDNY's Preparedness," 2002. Retrieved from http://www.nyc.gov/html/fdny/html/mck_report/toc.shtml.

MEMRI. "Report: Al-Qaeda Maghreb Commander Turns Self In," 2008. Retrieved from http://www.thememriblog.org/blog_personal/en/7847.htm

Metzinger, Thomas. *The Ego Tunnel: The Science of the Mind and the Myth of the Self.* New York: Basic Books, 2009.

Mind Hacks. "Erotic Self-stimulation and Brain Implants," September 16, 2008. Retrieved from http://mindhacks.com/2008/09/16/erotic-self-stimulation-and-brain-implants/.

Mithen, Steven. *After the Ice: A Global Human History 20,000–5,000BC.* Cambridge, MA: Harvard University Press, 2006.

Moyers, Bill. "Transcript: September 14, 2007." PBS, 2007. Retrieved from http://www.pbs.org/moyers/journal/09142007/transcript1.html.

Mydans, Seth. "Nasir Abas, Terrorist Defector, Aids Indonesian Police." *The New York Times,* February 29, 2008.

NASA. "Mission Information: STS-121," 2006. Retrieved from http://www.nasa.gov/mission_pages/shuttle/shuttlemissions/sts121/main/index.html.

National Geographic. "Killer Whale vs. Sea Lions," 2007. Retrieved from http://www.youtube.com/watch?v=DWsN63PRCW8.

Neisser, U. "Rising Scores on Intelligence Tests." *American Scientist,* 85, 440–7, 1997.
New York City Fire Department. "History of Fire Service," 2011. Retrieved from http://www.nyc.gov/html/fdny/html/history/fire_service.shtml.
NIST. "Final Report on the Collapse of World Trade Center Building 7," 2008. Retrieved from http://wtc.nist.gov/NCSTAR1/PDF/NCSTAR%201A.pdf.
Nuttin, J.M. "Narcissism beyond Gestalt and Awareness: The Name Letter Effect." *European Journal of Social Psychology,* 15(3), 353–61, 1985.

O'Keefe, Rosemarie. Audio: "Former NYC Commissioner Discusses Support Center for Families on 9/11," 2004. Retrieved from http://www.911memorial.org/blog/audio-former-nyccommissioner-discusses-support-center-families-911.
Olszewski, Karl. "Karl Olszewski a September 11th Hero," September 11, 2010. Retrieved from http://inchesawayfromsanity.blogspot.com/2007/09/karl-olszewski-september-11th-hero.html.
Operation Enduring Freedom. "iCasualties," 2011. Retrieved from http://www.icasualties.org/oef/.
Otten, Sabine and Dirk Wentura. "About the Impact of Automaticity in the Minimal Group Paradigm: Evidence from Affective Priming Tasks." *European Journal of Social Psychology,* 29 (8), 1049–71, 1999.

Panksepp, Jaak. *Affective Neuroscience: The Foundations of Human and Animal Emotions.* New York: Oxford University Press, 1998.
Perdue, C.W. et al. "'Us' and 'Them': Social Categorization and the Process of Intergroup Bias." *Journal of Personality and Social Psychology,* 59, 475–86, 1990.
Pogge, Richard W. "Real World Relativity: The GPS Navigation System," 2009. Retrieved from http://www.astronomy.ohio-state.edu/~pogge/Ast162/Unit5/gps.html.
Portnoy, Russell K. et al. "Compulsive Thalamic Self-stimulation." *Pain,* 27, 277–90, 1986.
Premack, D.G. and G. Woodruff. "Does the Chimpanzee Have a Theory of Mind?" *Behavioral and Brain Sciences,* 1, 515–26, 1978.

Qutb, Sayyid. *Milestones.* Chicago: Kazi Publications, 2003.

Ramachandran, V.S. "Mirror Neurons and Imitation Learning as the Driving Force behind 'The Great Leap Forward' in Human Evolution." Edge Foundation, Inc, (n.d.). Retrieved from http://www.edge.org/3rd_culture/ramachandran/ramachandran_p1.html.
Rasmussen, D. Tab, ed. *The Origin and Evolution of Humans and Humanness.* Boston, MA: Jones and Bartlett, 1993.
Richards, Michael P. et al. "Neanderthal Diet at Vindija and Neanderthal Predation: The Evidence from Stable

Isotopes." *Proceedings of the National Academy of Sciences,* 97 (13): 7663–66, June 20, 2000.

Ridley, Matt. *The Rational Optimist: How Prosperity Evolves.* New York: Harper, 2010.

Rigden, John S. *Einstein 1905: The Standard of Greatness.* Cambridge, MA: Harvard University Press, 2006.

RITA. "Bureau of Transportation Statistics," 2003. Retrieved from http://www.bts.gov/press_releases/2003/bts023_03/html/bts023_03.html and http://www.bts.gov/xml/air_traffic/src/datadisp.xml.

Robertson, Pat. "Pat Robertson: Islam Preaches Violence." *The Turkish Times,* 2002. Retrieved from http://www.theturkishtimes.com/archive/02/10_15/f_falwell.html.

Robertson, Pat. *The 700 Club,* November 9, 2009. Retrieved from http://mediamatters.org/mmtv/200911090042.

Rosenberg, Matt. "Current World Population," 2011. Retrieved from http://geography.about.com/od/obtainpopulationdata/a/worldpopulation.htm.

Sapolsky, Robert. "14. Limbic System," 2010. Stanford University lecture retrieved from http://www.youtube.com/watch?v=CAOnSbDSaOw.

Schachter, S. and J. Singer. "Cognitive, Social, and Physiological Determinants of Emotional State." *Psychological Review,* 69, 379–99, 1962.

Schmandt-Besserat, Denise. "Signs of Life." *Archaeology Odyssey,* January–February 2002.

Schneider, Mike. "Shuttle Crew Offloads Supplies for Space Station," August 20, 2006. Retrieved from http://www.katu.com/news/3633301.html.
Scola, Nancy. "Ghonim: 'Our Revolution Is Like Wikipedia.'" techPresident, 2011. Retrieved from http://techpresident.com/blog-entry/ghonim-our-revolution-wikipedia.
Semin, G.R. and A.S.R. Manstead. "The Social Implications of Embarrassment Displays and Restitution Behavior." *European Journal of Social Psychology,* 68, 441–54, 1982.

Semmelweis, Ignaz and K. Codell Carter, tr. *Etiology, Concept and Prophylaxis of Childbed Fever.* Madison, WI: University of Wisconsin Press, 1983.
The September 11 Digital Archive, 2004. Retrieved from http://911digitalarchive.org/.
Shaw, Thurstan, ed. *The Archaeology of Africa: Food, Metals, and Towns.* London: Routledge, 1995.
Shelov, Steven P. *Caring for Your Baby and Young Child, Birth to Age Five,* 5th ed. New York: Bantam Books, 2009.
Springer, Devin R et al. *Islamic Radicalism and Global Jihad.* Washington, DC: Georgetown University Press, 2009.
Stockwin, J.A.A. *Governing Japan: Divided Politics in a Resurgent Economy,* 3rd ed. New York: Wiley Blackwell, 1999.

Taylor, Heather. "Characteristics of New and First-time Home Buyers," September 1, 2010. Retrieved from http://www.nahb.org/generic.aspx?genericContentID=143996&fromGSA=1.

Thaler, Richard. "Toward a Positive Theory of Consumer Choice." *Journal of Economic Behavior and Organization,* 1, 39–60, 1980.

Thomas, Sean. "Gobekli Tepe: Your Questions Answered." *FirstPost,* 2006. Retrieved from http://www.thefirstpost.co.uk/2880,newscomment,news-politics,gobekli-your-questions-answered.

Tracy, Jessica and David Matsumoto. "The Spontaneous Expression of Pride and Shame: Evidence for Biologically Innate Nonverbal Displays." *Proceedings of the National Academy of Sciences,* 105 (33), 11655–60, August 11, 2008.

University of Utah. "Testosterone Levels Rise in Fans of Winning Teams," 1998. Retrieved from http://newswise.com/articles/testosterone-levels-rise-in-fans-of-winning-teams.

U.S. Department of Labor. "Bureau of Labor Statistics," 2010–2011. Retrieved from http://www.bls.gov/oco/cg/cgs041.htm.

U.S. Energy Information Administration. "Energy Units and
Calculators Explained," 2011. Retrieved from http://www.eia.doe.gov/energyexplained/index.cfm?page=about_btu.

Wade, Nicholas. *Before the Dawn: Recovering the Lost History of Our Ancestors.* New York: Penguin, 2006.

Wahab, Siraj. "Courageous US Woman Builds Bridges of Understanding." Arab News, 2006. Retrieved from http://www.mail-archive.com/proletar@yahoogroups.com/msg17229.html.

Wajda, Richard. BBC News, 2001. Retrieved from A href="http://news.bbc.co.uk/hi/english/static/in_depth/americas/2001/day_of_terror/eyewitness/default.stm"http://news.bbc.co.uk/hi/english/static/in_depth/americas/2001/day_of_terror/eyewitness/default.stm/.

Walker, Alan and Pat Shipman. *The Wisdom of the Bones.* New York: Knopf, 1996.

Wegner, D.M. *The Illusion of Conscious Will.* Cambridge, MA: MIT Press, 2002.

Wimmer, H. and J. Perner. "Beliefs about Beliefs: Representation and Constraining Function of Wrong Beliefs in Young Children's Understanding of Deception." *Cognition* 13 (1), 103–28, 1983.

Woolf, Alex. *Osama bin Laden.* Minneapolis, MN: Lerner Publishing, 2004.

Wright, Lawrence. *The Looming Tower: Al-Qaeda and the Road to 9/11.* New York: Knopf, 2006.

Wright, Lawrence. "The Rebellion Within: An Al-Qaeda Mastermind Questions Terrorism." *The New Yorker,* June 2, 2008.

Wyart, C. et al. "Smelling a Single Component of Male Sweat Alters Levels of Cortisol in Women." 27 (6), 1261–65, 2007.

about the authors and publisher

Peter Baumann began his career as a member of the internationally acclaimed 1970s band Tangerine Dream and later founded the Private Music record label. Instead of being derailed by early fame and fortune, he asked himself this in his late forties: "Given that I probably have about ten thousand days left on the planet, how can I use this time in the most meaningful and useful way possible?" To address this question, Baumann assembled a top-notch interdisciplinary think-tank, the San Francisco-based Baumann Foundation. He serves as a trustee of the California Institute of Integral Studies and as a fellow at the Mind & Life Institute.

Michael W. Taft is a serious student of evolution and the capacities of the human brain. A professional researcher and writer for more than two decades, Taft is fascinated by what neuroscience, biology, psychology, archaeology, and technology can tell us about the human condition. Michael is a senior adviser to the Baumann Foundation. He lives in Berkeley, California.

NE Press is dedicated to advancing understanding of the human condition.

front cover flap

"A human being is a part of a whole, called by us 'universe,' a part limited in time and space. He experiences himself, his thoughts and feelings, as something separated from the rest—a kind of optical delusion of his consciousness. This delusion is a kind of prison for us, restricting us to our personal desires and to affection for a few persons nearest to us. Our task must be to free ourselves from this prison..."

—Albert Einstein

The *self* we identify with—along with its biases and beliefs—turns out to be an evolutionary tool that made its first appearance some 50,000 years ago during what is called the conceptual revolution, arguably the most profound development in human history. The emerging ego accompanied our ability to construct more complex tools, create art, and redefine social structure. For the first time as a species, we were able to imagine the future, consider the thoughts of others, and picture ourselves in our own minds.

The emergence of the ego was an essential brain development necessary to create and function in the modern world. But, as the process of evolution continues—alongside the suffering caused by ego-identification—we increasingly recognize the ego's illusory nature and expand our awareness beyond its limitations.

back cover flap

Peter Baumann began his career as a member of the internationally acclaimed 1970s band Tangerine Dream and later founded the Private Music record label. Instead of being derailed by fame and fortune, he asked himself this in his late forties: "Given that I probably have about ten thousand days left on the planet, how can I use this time in the most meaningful and useful way possible?" To address this question, Baumann assembled a top-notch interdisciplinary think-tank, the San Francisco-based Baumann Foundation. He serves as a trustee of the California Institute of Integral Studies and as a fellow at the Mind & Life Institute.

Michael W. Taft is a serious student of evolution and the capacities of the human brain. A professional researcher and writer for more than two decades, Taft is fascinated by what neuroscience, biology, psychology, archaeology, and technology can tell us about the human condition. He has been working with the Baumann Foundation for the last several years as a program researcher. Michael lives in Berkeley, California.

back cover material

ego

"Baumann and Taft skillfully weave eyewitness accounts of 9/11 with insights from evolutionary theory, neuroscientific studies on brain plasticity and emotion, genetics, and other new areas of research. I highly recommend this book to all who are interested in how science can help in understanding both the human capacity for horrific action and the clear reasons for optimism about our collective future."

—Alfred W. Kaszniak, PhD, professor of psychology, neurology, and psychiatry, University of Arizona

"As evolution stumbles forward in its blind march of accidental brilliance, the radical insight that the ego is a mental self-concept will slowly become the prevailing viewpoint. Nothing special. It is the unfolding of the physical universe, the laws of nature, and evolution of life that are generating the expanded perspective that will allow humanity to find liberation from ego-identification."

—from *EGO*

A

60 Minutes, *130, 268*
Abbas, Nasir, *199*
ABC, *4, 193*
Accenture, *86*
adaptation, *282*
Afghanistan, *9-10, 95, 97-98, 108, 110, 128, 156-157, 171, 194, 199, 217, 227*
  war in, *9, 194*
afterlife, belief in, *12*
Age of Enlightenment, *272*
aggression, *74, 88*
aircraft, *187, 189*
Akihito, *167*
Aleppo, Syria, *214*
Alexandria, Egypt, *267*
Algeria, *199, 201*
alpha waves, *224*
al-Qaeda, *5, 9-10, 12-13, 106-108, 110-111, 169-171, 196, 199, 201-202, 204, 212-214, 216-218, 227, 264, 272*
  in Algeria, *199, 201*
  Hamburg terror cell, *10, 12, 110, 213-214, 217*
  ideology of, *12*
Al-Quds Mosque, *217*
altruism, *80, 82-83, 98, 149*
American Airlines, *13*
American Airlines Flight, *4-5, 9, 13, 25, 264*
American Board of Medical Specialties, *187*
American Medical Association, *187*
amygdala, *28-29, 47, 50*
anger, *29, 32, 48, 72, 74-75, 110*
animals, *164*
anterior superior temporal gyrus, *224*
anthropology, *35-36, 38, 138*
antidiuretic hormone (ADH), *46-47*
anxiety, *134-136, 236, 238*
  See also fear response,
anxiety attacks, *33*
Apollo 10, *189*
Al-Aqsa Mosque, *154*
Arab Spring, *266-269, 276*
Arafat, Yasser, *9*
archaeology, *164*
Argentina, *113*
Aristotle, *162-163, 202*
armies, *167*

Asia, *140-141*
atlatl, *140*
AT&T, *86*
Atta, Mohamed, *10, 110, 212-214, 216-218, 225, 233, 235, 264*
Augustus, *184*
Australopithecus, *116, 118, 123, 125, 182, 242*
authority, *166-167*
awareness, *201-202*
  conscious, *205-206, 208, 218, 220-222, 224, 231-232, 235, 238-239, 241-242, 246-247, 249-251, 253-254, 256-258, 260-261, 269, 272, 274-276, 278-280, 282-283*
  sensory, *257-258*
Axelrod, Steven, *32, 227*
Axial Age, *208*
Axial religions, *184, 231*
Azzam, Huthaifa, *199*
al-Fizazi, Iman, *217*
al-Hazmi, Nawaf, *110*
al-Mihdhar, Khalid, *110*
al-Sharif, Sayyid Imam ('Dr. Fadl'), *201*
al-Suqami, Satam, *13*
al-Zawahiri, Ayman, *171, 199, 201*

B

Bacon, Francis, *190*
Bali nightclub bombings, *106, 199*
Ban Ki-moon, *130*
Bank of America Tower, *110*
bees, *83*
behavior, *112-113*
behavior, learned, *112-113*
Beijing, China, *140-141*
belief, *131, 149, 280*
  See also religion,
belonging, *97-98*
  See also identification; social connections,
Benedict, Pope, *76*
Benotman, Noman, *199*
Bereitschaftspotential (BP), *220-221*
Bern, Switzerland, *208*
Bhattacharya, Joy, *222, 224*
bias blind spot, *197*
bin Laden, Mohammed, *154*
bin Laden, Omar, *199, 264*
bin Laden, Osama, *9-10, 107-108, 110, 127, 130, 154-157, 169-171, 196, 199, 217, 225, 264*
biological change, *190*
  See also evolution,
bipedalism, *116*
black Americans, *231-232*
'Black Hawk Down' incident, *171*

bonding, *43, 45, 53-55, 126-127, 144, 149-150, 152, 175*
brain imaging, *222, 224*
brain stem, *46*
Brickman, Philip, *66, 68*
Buddha, *184, 238-239*
'Burn a Koran Day', *128, 130-131, 146, 152, 227*
Bush administration, *194*

C

Cairo, Egypt, *213-214, 266-267*
Campbell, Donald, *66, 68*
Cantor Fitzgerald, *40, 42*
capitalism, conscious, *275*
caring, *77, 79-80, 82-84, 86-88, 90-92, 94-95, 97-99, 101*
Carroll, Michael, *64-66*
catecholamine hormones, *13*
Catholicism, *14, 75-76*
cell phones, *268*
centeredness, *274*
cerebellum, *46, 48*
change, *157, 159-160, 162, 189-191*
  See also evolution; human development,
Chicago, Illinois, *110*
childbearing, *55*
childbirth, *51, 53*
chimpanzees, *118, 142, 144, 164*
China, *140-141, 167, 184*
Chinese medicine, *162-163*
choppers, *119, 123*
Christianity, *75*
Chung, Connie, *42*
Churchill, Winston, *60-61*
Church of Jesus Christ of Latter-day Saints, *130*
cities, formation of, *166-167*
civilization, *166-167, 178-179, 183-184, 186-187, 189, 214, 229*
cognitive biases, *196-198, 205-206, 228, 239, 254, 256*
cognitive distortions, *196-197, 205*
  See cognitive biases,
communication, *127*
  See specific forms of communication,
compassion, *16, 84*
  See also empathy,
concepts, *127, 160*
conceptual abilities, *136, 141, 150, 152, 190, 205-206, 216, 232, 249-251, 253*
conceptual mind, *216-218, 220, 260-261*
conceptual phase, *271-272, 278*
conceptual revolution, *128, 130-131, 133-136, 138, 140-142, 144-146, 148-150, 152, 159, 164, 180, 182-184,*

186-187, 189-191, 193-194, 196-199, 201-202, 204-206, 208, 235, 275, 278
Confucius, 184
congenital insensitivity to pain with anhidrosis (CIPA), 22, 24, 38, 51
conscious capitalism, 275
contrition, 87
cooperation, 80, 82-83, 127, 141, 145, 183, 275
cortisol, 74
Cro-Magnons, 183
Crozet Islands, 113, 115
cultural development, 274-275, 278
  See also civilization,
Czech Republic, 9

D

danger, 27-28, 31
Darwin, Charles, 29, 201
  The Expression of Emotions in Man and Animals, 29
'Day of Wrath', 267
'Days of Ignorance', 12
death, 25, 150, 280
  escape from, 60, 64
  fear of, 12, 27
decision making, 220-222, 224-225, 228
Deecke, Lüder, 220
Descartes, René, 163, 190
developmental theory, 244, 246
DHA,
  See docosahexaenoic acid (DHA),
Diaz, Charles 'Chuckie', 57, 59-60, 227, 263-264
disgust, 36, 38
docosahexaenoic acid (DHA), 140-141, 145
Dove World Outreach Center, 128, 130
DreamTV, 268

E

East River, 79
eating, 72
EEGs, 220, 222
egalitarianism, 276
ego, 17, 71, 159-160, 162-164, 166, 171-172, 178-179, 198, 202, 206, 208
  development of, 244
  enlightenment and, 238-239, 241-242, 249-251, 253-254, 256-258, 260-261, 263-264, 274, 278-280, 282-283
  the organism and, 216, 218, 220-221, 225, 227-228, 233, 235
  sensations and, 25

Egypt, *155, 167, 184, 213-214, 266-269, 276*
Egyptian revolution, *266-268, 276*
Ein el-Hilweh, Palestine, *9*
Einstein, Albert, *16-17, 56, 63-64, 112, 134, 178-179, 191, 208, 224, 228, 256, 279*
Ekman, Paul, *35-36, 38*
Elijah, *184*
e-mail, *268*
embarrassment, *88, 90, 92*
emotions, *60-61, 63-66, 68, 70-72, 74-76, 90-91, 110-111, 115, 127, 217, 235*
  See also specific, emotions conceptual revolution and, *150, 152, 160, 172*
  emotional nervous system, *40, 42-43, 45-48, 50-51, 53-56*
  enlightenment and, *254, 256*
  evolution and, *25, 27-28, 31, 33, 43, 45-46, 50-51, 54*
  negative, *68, 70-72, 75*
  purpose of, *18, 20-22, 24-25, 27-29, 31-33, 35-36, 38 39*
  social, *77, 79-80, 82-84, 86-88, 90-92, 94-95, 97-99, 101*
  study of, *35-36, 38*
empathy, *16, 84, 148-149*
Empire State Building, *110*
endorphins, *53-54*
endowment effect, *175, 177, 197*
energy, *28, 43, 45, 82-83, 87-88, 138, 140-141, 149*
energy frugality, *87-88*
Enkidu, *166*
enlightenment, *257*
  developing, *236, 238-239, 241-242, 244, 246-247, 249-251, 253-254, 256-258, 260-261, 263-264*
  enlightenment revolution, *235, 275*
  process of, *246-247, 249-251, 253-254, 256-258, 260*
  rise of an enlightened humanity, *266-269, 271-272, 274-276, 278-280, 282-283*
Enlightenment, Age of, *272*
epinephrine, *35*
estrogen, *74*
euphoria, *60-61, 64-66*
European Union, *278*
  See also specific countries,
evolution, *113, 115-116, 118-120, 122-123, 125-126, 186-187, 189-190, 206, 208*
  caring and, *82-83, 101*
  conceptual revolution and, *131, 133, 138, 140-141*

emotions and, *25, 27-28, 31, 33, 43, 45-46, 50-51, 54*
enlightenment and, *242, 244, 246, 269, 271, 279-280, 283*
happiness and, *63-64, 66, 72, 74-76*
the organism and, *213, 217-218, 220, 228-229, 231-233, 235*
evolutionary psychology, *148-149*
experience, *206*
el-Shazly, Mona, *268*

F

Facebook, *267*
facial expressions, *36, 38*
Faizabad, Afghanistan, *128*
Falwell, Jerry, *130*
fatwas, *171*
fear response, *13-14, 21, 27-29, 31, 33, 46-47, 72, 133-136, 146, 214, 216*
See also anxiety,
feedback, *24-25, 27, 35, 38, 48, 64, 66, 72, 118-119, 141, 152, 217, 229*
fetal development, *242*
Fire Department of New York (FDNY), *6, 95, 120, 122, 218*
fishing, *140-141*
flesh-eating bacteria, *134-135*
flexibility, *274*
Florence, Colorado, *111*
Florida, *128, 130, 212*
Florie, Michael, *95*
Flynn, James, *231-232*
Flynn Effect, *231-232*
fMRIs, *222*
Fore people, *36, 38*
fortresses, *167*
fossils, *125-127*
France, *8-9*
Frederick, Susan, *119-120, 122-123, 227, 263-264*
freedom, *283*
Freud, Sigmund, *253*
frontal cortex, *33*

G

Gainesville, Florida, *128, 130*
Gatorade, *86*
Gautama, Siddhartha, *208, 235*
genetics, *83*
Germany, *8, 213-214, 216-217*
Ghonim, Wael, *266-268*
Giaramita, Frank, *95*
Gilgamesh, *166*
Gillette, *86*
Giuliani, Rudolph, *122*
globalism, *278*
Göbeckli Tepe ('Potbelly Hill'), *136, 138, 140, 145, 152, 166, 183*
god-kings, *167*

Golf Digest, *86*
Google, *267*
gracilization, *183*
Graham, Billy, *130*
Graham, Franklin, *130*
Grand Mosque, *154*
gravity, theory of, *208*
grief, *43*
Grinde, Björn, *63*
grooming, *142, 144*
Groundhog Day, *60*
Ground Zero, *79, 92, 94-95, 97-98, 227*
See also World Trade Center,
group cohesion, *88, 276*
group identification, *174-175, 177-178*
group survival, *80, 82*
Guantanamo Bay, Cuba, *111*
guilt, *88, 92*

H

Haeckel, Ernst, *242*
Hamburg, Germany, *217*
Hamburg terror cell, *10, 12, 110, 213-214, 217*
Hamm, Jon, *56*
hands, *116, 118-119*
happiness, *43, 45, 48, 57, 59-61, 63-66, 68, 70-72, 74-76*
evolution and, *63-64*
Harvey, William, *201*
hedonic adaptation, *66, 68*
hedonic treadmill, *66, 68*
Hélie, Sébastien, *222*
herd behavior, *80, 82*
Hernandez, Peter, *95*
heroism, *98*
Hess, Michael, *191, 193*
hijackers (September 11, 2001, terrorist attacks), *13-14, 25, 196, 212-213, 268*
See also Atta, Mohamed; September 11, 2001 terrorist attacks,
hippocampus, *47*
Hitler, Adolf, *130*
homeostasis, *24-25, 46, 48*
Homo erectus, *123, 125-127, 138, 142, 145-146, 229, 232, 244*
Homo habilis, *118-119, 123*
Homo sapiens, *127, 138, 142, 145, 148, 232, 244*
hormones, *74*
human condition, *16, 264*
human development, *106-108, 110-113, 115-116, 118-120, 122-123, 125-127, 186-187, 189, 239, 241-242, 244, 246-247, 249-251, 253-254, 256-258, 260, 269, 271, 274-275, 282-283*
See also evolution,

human domestication, *182-183*
human rights, *274, 283*
hunger, *38*
hunter-gatherers, *4-5, 9-10, 87-88, 99, 125-126, 134, 136, 140, 166, 182, 214, 216*
  See also Paleolithic era,
Hussein, Saddam, *9, 169*
Hutton, James, *201*
hypothalamus, *46-47*

I

Iacoboni, Marco, *90-91*
ideas, *149, 152*
identification, *174-175, 177-178, 218, 235, 241, 251, 253-254, 256-258, 260-261, 263, 272, 274-276, 278-280, 282-283*
identity, *159-160, 162*
illusions, optical, *197-198*
imagination, *14, 17, 110-112, 115, 119, 127, 136, 148-149, 152, 159, 164, 214, 216*
implicit association test (IAT), *174-175, 206*
implicit priming, *174*
Independence Day, *21*
India, *128, 184*
Indonesia, *128*
infants, *164*
insects, *83*
intelligence, *115, 231-232*
interconnectedness, *45, 276*
  See also social connections,
International Court of Justice, *274*
International Humanist and Ethical Union, *130*
International Space Station, *8, 180*
international trade, *278*
IQ tests, *231-232*
Iraq, *9, 169, 194*
Iraq War, *9, 194*
Isaiah, *184*
Islam, *10, 12, 128, 130-131, 133, 155-157, 171, 214, 217, 225, 227*
  See also Muslims,
Israel, *108, 110, 217*

J

jahililyya, *12*
James, William, *32-33, 236, 238, 264*
  Principles of Psychology, *238*
Japan, *167*
jealousy, *182*
Jeddah, Saudi Arabia, *154-156, 199*

Jeddah Economic Forum, *264*
jellyfish, *133*
Jennings, Barry, *191, 193-194, 204*
Jericho, Palestine, *167, 183*
Jewish people, *12, 170, 217-218*
jihad, *155, 199, 201*
Jobs, Steve, *275*
Jones, Terry, *128, 130-131, 146, 227, 264*
'Islam Is of the Devil', *128*
joy, *43, 45*
Jung-Beeman, Mark, *222, 224*

K

Kabul, Afghanistan, *128*
Kaczinski, Ted, *111*
Kandahar, Afganistan, *108, 110*
Khartoum, Sudan, *170*
King Abdulaziz University, *155*
kin selection, *83*
Kitty Hawk, North Carolina, *187*
Klett, David, *106*
Kornhuber, Hans Helmut, *220*
Kounious, John, *222, 224*
Kreider, Tim, *60, 64*
Kronos, *8*
Kurzweil, Ray, *189*
Kuwait, *169*

L

Lady Gaga, *56*
language, *136, 141-142, 144-145, 148-149, 152, 183-184*
leadership, *166-167, 276*
Leakey, Richard, *126-127*
learned behavior, *112-113*
LeDoux, Joseph, *27-28, 163, 214*
Leeuwenhoek, Antonie van, *201*
Le Monde, *8-9*
Leonardo module, *180*
Lewin, Daniel, *13*
Libet, Benjamin, *220-221*
Library Tower, *110*
Libyan Islamic Fighting Group, *199*
limbic system, *46*
Lohan, Lindsay, *87*
Loose Change, *194*
Los Angeles, California, *110*
loss, *43*
love, *48, 51, 53-55*
'Lucy', *116*
Luna, *189*
Lutnick, Edie, *56*
Lutnick, Gary, *40, 42-43*

Lutnick, Howard, *40, 42-43, 227, 263-264*
Lykken, David, *61, 63*

M

Mahavira, *184*
mammals, *29, 45, 66*
Manstead, Anthony, *88, 90*
manual dexterity, *118-119*
marriages, mixed-race, *278*
martyrdom, *12*
Maslow, Abraham, *61*
mass media, *56, 134-135*
mathematics, *184*
mating, *31*
Mead, Margaret, *36*
Mecca, *199*
medicine, *186-187*
meditation, *257*
Medvedev, Dmitry, *275*
memes, *149*
memory, *110, 115, 206*
Memphis, Egypt, *184*
men, *74*
Menes, *184*
Merkel, Stephen, *42*
Mesoamerica, *184*
Metzinger, Thomas, *198*
military power, *167*
Millenium Plot, *106*
Miller, Earl, *224*
mindfulness meditation, *257*
mirror neurons, *90-91*
mirror stage, *164*
modernity, *216-217, 233*
Mohammed (the Prophet), *130, 169*
Mohammed, Khalid Sheikh, *106-108, 110-111, 127, 217, 225, 264*
motivation, *51, 61, 111, 217*
Moussaoui, Zacarias, *111*
Mubarak, Hosni, *171, 266, 269*
mujahedin, *169-170*
multiculturalism, *278*
Murray, Bill, *60*
Muslims, *12, 110, 128, 130-131, 133-134, 146, 155, 157, 169, 214, 217, 225, 227*
  See also Islam,

N

Nablus, Palestine, *9*
Nantucket Inquirer and Mirror, *32*
National Enquirer, *84*
national identification, *272, 278*
National Institute of Standards and Technology, *194, 202*
National Police Day, *267*

NATO (North Atlantic Treaty Organization), *128, 272*
natural selection, *16, 27-28, 31, 87, 101, 118, 125-126, 133*
Neanderthals, *138, 140, 142, 145, 244*
negativity bias, *133-134, 136, 196-197*
neocortex, *46*
nervous system, *38, 40, 42-43, 45-48, 50-51, 53-56, 65-66, 71, 101, 112-113, 133-134, 144, 233*
  emotional, *22, 24-25, 27-29, 31-33*
Network, *74*
neural baseline, *66, 68, 70*
neurons, *50, 90-91, 112-113, 233, 235*
neuroplasticity, *61, 63*
neuroscience, *12, 21-22, 27-29, 33, 38, 45, 50, 163, 220-221, 233, 235, 244*
  See also brains,
Newton, Isaac, *201, 208*
New York, New York,
  See September 11, 2001 terrorist attacks,
New York City Housing Authority, Emergency Services Department, *191*
New York Police Department (NYPD), *6, 8, 57, 59-60, 95, 122, 218*
New York Times, *21, 60*
Nichols, Terry, *111*
Nile river, *184*
Ninlil, *166*
non-zero sum situations, *83-84*
Nordegren, Elin, *84, 88, 146*
normalcy bias, *197*
Norway, *9*
Nowak, Lisa, *88, 180, 182*
NYU (New York University) hospital, *77*

O

Obama, Barack, *130*
Oefelein, William, *182*
Office of Emergency Management, *122, 191*
O'Keefe, Rosemarie, *77, 79-80*
Olszewski, Karl, *94-95, 97-98, 227, 263-264*
Olympics, *98-99*
Omega-3 fats, *141*
Ong, Betty, *13*
OpenLeaks, *276*
openness, *274-276*
opiates, *43*
orangutans, *164*
orcas, *113, 115*
oxytocin, *53, 74*

P

pain, *22, 24-25, 28, 38, 43, 48, 50-51, 66, 71, 101, 163*
pair-bonding, *126*
Pakistan, *128, 156*
Paleolithic era, *4-6, 8-10, 21, 43, 45, 53, 63, 74, 80, 82, 87-88, 99, 131, 133, 214, 227*
Palestine, *9, 167, 183*
Palestinians, *9*
Pan American, *138, 189*
Paralympics, *98-99*
parasympathetic nervous system, *144*
partisanism, *278*
passions, *55-56*
Patagonia, *275*
Pearl, Daniel, *106*
perception, *196*
perceptions, *238-239*
personal phase, *246, 249-251, 253, 256*
Peshawar, Pakistan, *156*
Phase Model of Development, *246-247, 249-251, 253-254, 256-258, 260, 264, 269, 271*
Phase Model of Human Evolution, *269, 271*
Philippines, *107*
philosophy, *55-56*
Phoenicians, *189*
Pier, *75, 79*
piloerection, *31*
planning, *115, 123, 127, 136, 159-160*
pleasure, *22, 24-25, 28, 45, 48, 50-51, 71, 101, 163*
politics, *278-279*
Pompano Beach, Florida, *212*
Pony Express, *187*
Port Authority (PA), *92, 94-95, 97-98, 227*
positive bonding, *43*
post-conceptual orientation, *261, 263, 278-279*
post-conceptual phase, *272, 274*
post-personal phase, *246, 256-258, 260-261, 263-264, 274-275, 280*
pre-conceptual phase, *271, 278*
prediction, *115, 136*
prefrontal cortex (PFC), *47-48, 224*
pre-personal phase, *246-247, 249, 282-283*
pride, *98-99*
primates,
See specific primates,
prison of feelings, *4-6, 8-10, 12-14, 16-17*
Prophet's Mosque, *154*

psychology, *32-33, 236, 238, 242, 244*
Puopolo, Dominic, *13*
Puopolo, Sonia, *13-14, 25, 264*
Puopolo, Tita, *264*
Puyi, *167*
pyramids, *167*

Q

Qutb, Mohammed, *155*
Qutb, Sayyid, *12, 155*

R

rabbits, *113*
race, *231-232*
racism, *175*
Rahsidiyeh, Palestine, *9*
railways, *187*
Ramachandran, V.S., *91*
rats, *27*
recapitulation theory, *242*
Red Cross, *274*
regret, *88, 92*
Reid, Richard, *111*
rejection, *97*
relativity, theory of, *208*
relaxation, *222*
religion, *14, 17, 55-56, 75-76, 128, 130-131, 148-150, 152, 166-167, 184, 231, 258, 280*
  See also specific religions,
reproduction, *51, 53, 55, 76*
reptiles, *45*
reptilian brain, *45-46, 214*
Revolution 2.0, *268*
Robertson, Pat, *130*
Roman Catholic Church, *76*
Roman empire, *184, 190*
Roughan, Christopher, *95*
Russia, *275*
  See also Soviet Union,

S

sadness, *43*
Safe Horizons, *79*
Said, Khaled, *267, 269*
Salazar, Roberto, *22, 24, 38, 51, 66*
Saudi-American Exchange Program, *264*
Saudi Arabia, *169-171, 199*
Saudi Binladen Company, *154*
Sayyaf, Rasul, *107*
scientific revolution, *201-202, 205*
seals, *113, 115*
Sears Tower, *110*
Seattle, Washington, *110*
selective perception, *196*

self-concept, *159-160, 162-164, 171-172, 178, 198, 206, 218, 227-228, 233, 235, 241, 249-251, 253, 258, 274, 279-280, 282-283*
self-reflection, *164, 231, 254*
self-serving bias, *197*
Semin, Gun, *88, 90*
Semmelweis, Ignaz, *186, 190*
Semmelweis effect, *197*
sensations, *24-25, 38, 50-51, 233, 235*
  See also pain; pleasure,
septal nuclei, *50*
September 11, 2001, terrorist attacks, *4-6, 8-10, 12-14, 32-33, 40, 84, 92, 94-95, 97-98, 106-108, 110-111, 119-120, 122-123, 127-128, 130-131, 133, 140, 146, 191, 193-194, 196-197, 199, 202, 204, 213, 217-218, 268*
  international reaction to the, *8-9*
  victims of the, *13-14, 18, 20-21, 25, 27, 39-40, 42-43, 56-57, 59-60, 77, 79-80, 94-95, 97-98, 119-120, 122-123, 191, 193-194, 204, 227, 263-264*
September 11 Digital Archive, *8*
serotonin, *54*
Seven Deadly Sins, *75-76*
Seven World Trade Center, *191, 193-194, 202, 204*
sex abuse scandal, *76*
sexuality, *31, 51, 53-55*
shamanism, *150*
shame, *88, 90, 92, 98*
Shipman, Colleen, *182*
'shoe bomber' plot, *106*
Siwik-Daniels, Veronica, *86*
slavery, *283*
Smith, Adam, *201*
social connections, *43, 45, 47, 80, 82-83, 127, 149-150, 152, 276*
social emotions, *77, 79-80, 82-84, 86-88, 90-92, 94-95, 97-99, 101*
social media, *267-269*
social organization, *189, 229, 232-233, 274, 276*
social responsibility, *276*
Socrates, *184, 221*
Somalia, *128, 171*
sorrow, *43*
South Korea, *9*
Soviet Union, *156-157, 169, 171, 227, 272*
spacecraft, *187, 189*
Space Imaging, *8*
Space Needle, *110*
Space Shuttle Discovery, *180*
space travel, *187, 189*
speech, *144-145*
  See also language,

Spencer, Herbert, *242*
squirrel monkeys, *112*
stakeholdes, *275*
Start Something program, *86*
State Security Investigations Service, *266-267, 269*
Stevens, Wallace, *260*
St. Jude Hospital, *86*
Stone Age,
  See Paleolithic era,
strep, *134-135*
subconscious decision making, *221-222, 224-225, 228*
submission, *88*
Sudan, *170-171*
Sun, Ron, *222*
Sunni Islam, *217*
superbugs, *134-135*
superiority bias, *197*
survival mechanisms, *10, 12, 27-29, 51, 216*
Sweeney, Madeline, *13*
Switzerland, *206, 208*
sympathetic nervous system, *13*
Syria, *155*

T

The 700 Club, *130*
1993 World Trade Center bombings, *106-107, 122*
2001: A Space Odyssey, *10*
the Bible, *202*
the body, emotions and, *32-33, 35-36, 38-39*
'Bojinka', *107-108*
the brain, *12, 14, 17, 21-22, 24-25, 32-33, 45-47, 112-113, 116, 163, 198, 220, 222, 224-225, 280*
  See also neuroscience; specific parts of the brain,
  development of, *118-119, 123, 241*
  emotional, *46-47*
  evolution of, *50-51, 131, 133, 138, 140-141, 186-187, 189, 229, 231-233, 235, 280*
  mammalian, *46*
  Paleolithic, *63*
the Capitol, *110*
The Container Store, *275*
the Internet, *267-269, 272*
the Middle East, *184*
  See also specific countries,
the organism, *212-214, 216-218, 220-222, 224-225, 227-229, 231-233, 235, 238, 258, 261, 263, 279-280, 282*
the Pentagon, *4, 110*

the Qur'an, *130-131, 146, 152, 227, 235, 264*
TAG Heuer, *86*
Tahrir Square, *266*
Taliban, *9, 171*
Target House, St. Jude Hospital, *86*
technology, *184, 186-187, 189-190, 268-269, 274, 276*
teeth, baring of, *29*
Tenno, *167*
terrorist attacks of September 11, 2001.,
See September 11, 2001 terrorist attacks,
terrorists, *12*
See also al-Qaeda; hijackers (September 11, 2001, terrorist attacks); specific terrorists,
testosterone, *74*
text messaging, *268*
thalamus, *22, 24*
theocracy, *167*
theory of mind, *146, 148-150*
the 'Pit', *92, 94-95, 97-98*
Tianyuan Cave, *140-141*
Tiger Woods Foundation, *86*
Tiger Woods Learning Center, *86*
tolerance, *274*
tools, *118-119, 123, 127, 136, 138, 140-141, 229, 232-233*
Tora Bora, Afghanistan, *156*
trade, international, *278*
transgressions, *87-88*
transparency, *276*
transportation, *186-187, 189*
See also specific forms,
travel, *278*
See also space travel; transportation,
tribal bonds, *175*
trust, *149*
Turkana Boy, *126-127*
Turkey, *9, 136*
Twin Towers, *4-6, 8-10, 18, 20, 40, 57, 59-60, 80, 84, 107, 119-120, 122-123, 191, 193-194, 196-197, 202, 204*
See also World,
Trade Center Twitter, *267-268, 275*
the Web, *272, 276*
the West, *216-217, 233*

U

United Arab Emirates, *267*
United Nations, *130, 274*
United States, *4-6, 8-10, 12-14, 16-17, 169-171, 217-218*
See also September 11, 2001, terrorist attacks,

Upanishads, *208*
Ur, *184*
Uruk, *166, 183*
U.S. Air Force, *187*
U.S. Naval War College, *122*

V

Valdes Peninsula, *113*
Valium, *33, 35*
vasopressin, *53, 74*
verification, *222*
Vesalius, Andreas, *201*
violence, *75, 88, 111, 182-183, 239, 278*
  See also terrorism,

W

Wade, Nicholas, *229*
Wajda, Richard, *18, 20-21, 27, 39, 56, 227, 263-264*
Wallas, George, *221-222*
warfare, *9-10*
Warrick, Brad, *212-213*
Warrick's Rent-a-Car, *212, 217*
Washington, DC.,
  See September 11, 2001 terrorist attacks terrorists,
whales, *113, 115*
White House, *110*
Whole Foods Market, *275*
WikiLeaks, *276*
Wikipedia, *268*
Wilde, Oscar,
  The Picture of Dorian Gray, *75*
Wilson, Stephanie, *180*
women, *51, 53, 55, 74*
Woods, Tiger, *84, 86-88, 146*
World Evangelical Alliance, *130*
'World Islamic Front for Jihad Against Jews and Crusaders', *171*
World Trade Center, *4-6, 8-9, 18, 20, 32-33, 39-40, 57, 59-60, 77, 80, 84, 92, 94-95, 97-98, 106-108, 110-111, 119-120, 122-123, 127, 131, 133, 191, 193-194, 196-197, 202, 204, 213, 217-218*
  See also September 11, 2001, terrorist attacks,
worst-case scenario, *133-134*
Wright brothers, *187*
writing, *183-184*
  See also language,

Y

Yemen, *154, 170-171*
Young, Thomas, *276*
Yousef, Ramzi, *107-108, 110-111*
YouTube, *268*

Z

Zappos, *275*
Zoroaster, *184*

Books For ALL Kinds of Readers

At ReadHowYouWant we understand that one size does not fit all types of readers. Our innovative, patent pending technology allows us to design new formats to make reading easier and more enjoyable for you. This helps improve your speed of reading and your comprehension. Our EasyRead printed books have been optimized to improve word recognition, ease eye tracking by adjusting word and line spacing as well as minimizing hyphenation. Our EasyRead SuperLarge editions have been developed to make reading easier and more accessible for vision-impaired readers. We offer Braille and DAISY formats of our books and all popular E-Book formats.

We are continually introducing new formats based upon research and reader preferences. Visit our web-site to see all of our formats and learn how you can Personalize our books for yourself or as gifts. Sign up to Become A RHYW Registered Reader.

www.readhowyouwant.com

Made in United States
Orlando, FL
04 November 2026

90202664R00195